Planet Earth
A Beginner's Guide

"Here really is everything you might want to know about the bowels of the Earth – and inevitably about plate tectonics, the atmosphere, and more ... An intimate exploration of this unusual (at least in our solar system) and beautiful planet."

Brian Clegg – author of *Inflight Science* and
Before the Big Bang

"An interesting and easily readable overview of the history and evolution of planet Earth and life upon it, with due credit given to the historical contributions of important scientific personalities."

William Lowrie – Emeritus Professor of Geophysics
at the Swiss Federal Institute of Technology

"Concisely, authoritatively, and very clearly, John Gribbin has produced an engaging and very up-to-date picture of how the Earth and its organisms have developed over time. A wonderful narrative."

Andrew Goudie – Emeritus Professor in Geography
at the University of Oxford

"John Gribbin has done it again! Another first-rate account of a subject of immense importance, and beautifully written. From the fires within to the icy poles, the restless continents and dynamic atmosphere, this is a superb introduction to the inner and outer workings of the planet we call home."

Dr. Lewis Dartnell – University College London

ONEWORLD BEGINNER'S GUIDES combine an original, inventive, and engaging approach with expert analysis on subjects ranging from art and history to religion and politics, and everything in between. Innovative and affordable, books in the series are perfect for anyone curious about the way the world works and the big ideas of our time.

Planet Earth
A Beginner's Guide

John Gribbin
with
Mary Gribbin

ONEWORLD

A Oneworld Paperback Original

Published by Oneworld Publications 2012

Copyright © John and Mary Gribbin 2012

ISBN 978-1-85168-828-9

Typeset by Cenveo Publisher Services, Bangalore, India
Cover design by vaguelymemorable.com

Printed and bound in Great Britain by
CPI Group (UK) Ltd, Croydon, CR0 4YY

Oneworld Publications
185 Banbury Road
Oxford OX2 7AR
England

Learn more about Oneworld. Join our mailing list to
find out about our latest titles and special offers at:
www.oneworld-publications.com

Contents

Foreword
Home planet

We live on a planet. Our home is a nearly spherical ball of rock, about 12,700 kilometres (7900 miles) in diameter, covered with a thin smear of water and gas, orbiting around the star we call the sun once every year. As it orbits, the Earth turns on its axis, at present once every twenty-four hours, giving us the cycle of day and night. This has been going on for, in round numbers, 4.5 billion (thousand million) years, the time since the sun and its family of planets formed. That is the extent of the history of terrestrial time, the framework within which the present features of the Earth, including its suitability as a home for life, have evolved.

We cannot tell that whole story here. But we will tell you how the forces that shaped the Earth are still active today, causing earthquakes and volcanoes, ripping the Earth's crust apart as ocean basins grow, and pushing landmasses around in a changing pattern that produces collisions between continents in which great mountain ranges grow. Both geological activity and life combine to influence the composition of the air that we breathe, and life itself both responds to changes in the climate of our planet and drives some of those changes, including the present warming of the globe.

The story of our home planet as it is today is not just of interest to Earth scientists, but is of compelling importance to everyone alive on Earth today, at the mercy of inexorable geological forces. Now, let's take our journey.

1

A brief history of terrestrial time

Before we begin the story of our home planet, we should first consider another story of terrestrial time, the story of how pioneering geologists came to realize the true extent of the history of the Earth, and the nature of the forces that have shaped, and continue to shape, the world we live in. That history goes back scarcely more than two hundred years, to the time when scientists in Europe first began seriously to question the idea, derived from a literal interpretation of biblical stories, that the Earth is only some six thousand years old.

The birth of geology

The modern understanding of the Earth began with the work of the Scotsman James Hutton, who had qualified as a physician, but was wealthy enough never to have to work and devoted his life to science. He presented his ideas to the Royal Society of Edinburgh in 1785, and published them in a book, *Theory of the Earth*, in 1795. The Geological Society of London, the first organization devoted to the study of what are now called the Earth sciences, was founded just twelve years later, in 1807. By then, Hutton's pioneering ideas had been popularized by his friend John Playfair in his book *Illustrations of the Huttonian Theory of the Earth*.

At the end of the eighteenth century, although it was still widely believed that the Earth had been created in 4004 BC, scientists like Hutton were well aware that our plant must be much older than this. The question was, how old? And how had the landscape been shaped into its present form?

The puzzle had been hinted at as far back as 1665, by Robert Hooke, in his book *Micrographia*. Hooke was one of the founding fathers of science, but doesn't always get the credit he deserves because he was a contemporary of Isaac Newton, compared with whom even the cleverest people look ordinary. Hooke was one of the first people to realize that fossils aren't just curiously shaped rocks that happen to resemble living things, but are the preserved remains of living things themselves. Describing ammonites in his book, he wrote that they are 'the Shells of certain Shelfishes, which, either by some Deluge, Inundation, Earthquake, or some such other means, came to be thrown to that place, and there to be filled with some kind of mud or clay, or *petrifying* water, or some other substance, which in tract of time has been settled together and hardened'. In lectures at Gresham College, in London, he also said that 'Parts [of the Earth] which have been sea are now land', and that 'mountains have been turned into plains, and plains into mountains, and the like'. He was exactly right on every count. But few people took any notice of these ideas at the time, because such dramatic changes clearly could not have happened in the span of a few thousand years.

Exactly at the same time, in the mid 1660s, the Dane Niels Steensen also recognized that fossils found far inland today, and even on high mountains, are the remains of creatures that once lived in the sea. He wrote under a Latinized version of his name, as Steno, and said that different rock layers ('strata') must have been laid down underwater at different times in the past, as a result of floods covering the land. So the idea of fossils as the remains of living creatures became linked with the idea of the biblical Flood, which some people saw as just the latest in a series of watery cataclysms that had inundated the Earth. This was quite different from Hooke's idea that the land itself could be raised up or lowered, while sea levels stayed the same. But it still left the question of how long all this would have taken – and where all the water came from.

In the 1740s, Carl Linnaeus, remembered today for his invention of a classification system for plants and animals, came up with his own version of the Flood idea. He realized that even if it had really happened, the Flood described in the Bible had not lasted long enough for all these processes to have taken place. Instead, he thought that the Earth had originally been completely covered with water, which has been gradually receding ever since, leaving behind fossils of sea creatures as dry land emerged from beneath the waves. The whole point of this idea was that it would have taken much longer than the six thousand years of Earth history suggested by adding up chronologies in the Bible, but Linnaeus was careful not to get into trouble with the Church by actually saying as much in print. The most daring comment he allowed himself to make for public consumption appeared in 1753, when, in his book *Museum Tessinianum*, he wrote, 'The infinite number of fossils of strange and unknown animals buried in the rock strata beneath the highest mountains, animals that no man of our age has beheld, are the only evidence of the inhabitants of our ancient earth at a period too remote for any historian to trace.'

But another scientist didn't exercise such restraint. While Linnaeus was in Sweden thinking about the implications of fossils for the terrestrial timescale, his contemporary the Comte de Buffon was in France carrying out the first actual experiments to try to determine the age of the Earth. Buffon was a hard-working and successful scientist, even though he lived off inherited wealth and could have spent his days in idle luxury. He was so dedicated that he actually hired a servant whose job was to drag Buffon out of bed each morning at 5 a.m. and make sure he was awake and ready to start work. His great achievement was to produce a huge survey of science, the *Histoire naturelle*, in no less than forty-four volumes; his most inspired piece of work is described in the book.

Buffon thought that the Earth had been formed from a blob of molten material torn out of the sun by a passing comet, which wasn't a completely crazy idea in the eighteenth century. So he tried to work out how long it would take for a ball of molten iron the size of the Earth to cool down to its present temperature. He did this by heating balls of iron of different sizes until they were red hot and on the point of melting, then timing how long it took them to cool down to the point where they could just be touched without burning the skin. The story goes that his assistants in this experiment were aristocratic ladies with delicate hands, protected by the finest silk gloves, who he used, in effect, as thermometers – there were, of course, no accurate thermometers that could do the job in his day.

Scaling the results up from his experiments to a ball the size of the Earth, Buffon calculated that the Earth must be at least seventy-five thousand years old, and dared to say as much in the *Histoire*. Even though this age is so much less than the billions of years suggested by modern science, this was a landmark event. Buffon had dared to publish an age for the Earth more than ten times greater than the age suggested by Bible scholars. The next person to make such an estimate pushed the date back even farther – but, unlike Buffon, lacked the courage to publicize his findings.

Joseph Fourier is remembered today as a mathematician; but he developed the mathematical techniques for which he became famous for a practical reason – to describe the way heat flows from a hot object to a colder one. The equations can describe, for example, the rate at which heat flows along an iron bar if one end is kept red hot in a furnace and the other end is at room temperature. Fourier used these equations to estimate the rate at which the Earth would have cooled down from a molten state, and, unlike Buffon, he also realized that once a crust of solid rock had formed at the surface of the Earth it would act like an insulating blanket and slow down the rate at which heat was being lost to space. Putting everything together, he came up with a formula which, when solved, gives the age of the Earth as 100 *million* years. Fourier was so shocked by this that he never published that number. He did publish the formula, in 1820, and any half-decent mathematician could have used it (as they surely must have) to work the number out. But Fourier just couldn't bring himself to stir up controversy by suggesting that the Earth was nearly a hundred thousand times older than Bible scholars believed. By 1820, though, this kind of timescale was, if anything, too short for the requirements of geologists following in the footsteps of James Hutton.

The uniformitarians

The key thing that Hutton appreciated was that there is no need to invoke global catastrophes to explain how the Earth got to be the way it is today. He studied the way strata that must have originally been laid down one on top of the other are now seen to be bent and twisted into distorted patterns, and appreciated that instead of explaining these features as due to a global catastrophic upheaval, they could be caused by the same processes we see at work on Earth today, but operating over immensely long periods of time. These processes include volcanic eruptions and great earthquakes, which are catastrophic enough for anyone living in their vicinity. But, Hutton realized, such short-lived events are common on a geological timescale, and as normal in the life of the Earth as sneezing is in the life of a human being. The idea that gradual processes operating over immense periods of time can explain the origin of all the features of the Earth, from mountain ranges to ocean basins, became known as uniformitarianism, since the same uniform processes we see at work today can explain everything from how mountain ranges are uplifted from the sea floor to the way they are worn away by erosion, with the resulting sediments being laid down under the sea, eventually to become the raw material of new mountain ranges. This image of the changing Earth neatly explains why there are just three kinds of rock on Earth – igneous rock, which flowed from volcanoes in a molten state and set hard; sedimentary rock, laid down underwater from tiny pieces of older rock worn away by erosion; and metamorphic rock, such as granite, which is formed when either of the two basic kinds of rock becomes at least partially molten and reworked.

Hutton couldn't calculate the age of the Earth, but he understood that all these processes need a very long time to operate. Indeed, he thought that there might not have been a 'beginning' at all, and that the Earth had always existed, and always would exist, with these processes going on forever. In a paper published in 1788, he wrote, 'The result, therefore, of our present enquiry is, that we find no vestige of a beginning – no prospect of an end.'

The person who built on Hutton's foundations and really made people take notice of the idea of uniformitarianism was another Scot, Charles Lyell, who published a great book, *Principles of Geology*, in three volumes, between 1830 and 1833. In the typical style of the day, he provided a subtitle which explained what the book was all about: 'Being an Attempt to Explain the Former Changes of the Earth's Surface by Reference to Causes Now in Operation'. Among the evidence gathered in his book, Lyell described how at Mount Etna, layers of igneous rock formed by lava flows are separated by layers of sedimentary rock. In one place, a bed containing fossilized oysters 'no less than *twenty feet in thickness*, is there seen resting on a current of basaltic lava; upon the oyster bed again is superimposed a second mass of lava'. So the time interval between the lava flows was long enough for sediments twenty feet thick to be laid down.

And the lava beds themselves weren't laid down overnight. Lyell calculated that it would require 'ninety flows of lava, each a mile in breadth at their termination, to raise the present foot of the volcano as much as the average height of one lava-current'.

Among the people on whom Lyell's book made a deep impression was the young Charles Darwin, who set off on his famous voyage on the *Beagle* in 1831, taking the first volume of *Principles of Geology* with him – the other volumes caught up with him on his travels. Over the years that followed, as Darwin developed his ideas about the origin of species as a result of evolution by natural selection, he grasped that this, too, is a uniformitarian process requiring immense timescales. Thanks to Lyell (and to Hutton before him) Darwin knew that nature had indeed provided a sufficient timescale for natural selection to do its work. He said that Lyell had given him 'the gift of time', and much later commented, 'I always feel as if my books came half out of Lyell's brain ... the great merit of the Principles was that it altered the whole tone of one's mind.'

Darwin's book *On the Origin of Species* was published in 1859. Not everyone was an immediate convert to his ideas, but it is fair to say that by the last quarter of the nineteenth century many geologists and biologists were convinced that the Earth must be at least hundreds of millions of years old. The snag was, the physicists and astronomers were telling them that this was impossible, according to all the known laws of physics. Both sides of the argument were right. The conflict would only be resolved, and the dating story brought up to date, when previously unknown laws of physics were discovered.

Dating up to date

By the middle of the nineteenth century, physicists had realized that nothing lasts forever; everything wears out, eventually.

This key principle became enshrined in a scientific law, the famous second law of thermodynamics. It meant that there must indeed have been a 'vestige of a beginning' at some remote time in the past, and that, one day, albeit in the far future, there would be an end to the kind of conditions that exist on Earth today. So in 1852, the British physicist William Thomson (who later became Lord Kelvin, the name by which he is better known) wrote:

> Within a finite period of past time the earth must have been, and within a finite period of time to come the earth must again be unfit for the habitation of man as at present constituted, unless operations have been or are to be performed which are impossible under the laws to which the known operations going on at present in the material world are subject.

The Earth as we know it could not exist as a home for life without the sun, so Thomson tried to work out how long the sun could keep pouring out the vast quantities of heat and light that make the Earth habitable. Using an example with which his Victorian contemporaries would have been comfortable, Thomson pointed out that even if the sun were made entirely of coal, burning in a pure oxygen atmosphere, it would only last for a few thousand years before becoming a cinder. But he found that there is another source of energy that a star like the sun can draw on.

Thomson realized (as his German counterpart Hermann von Helmholtz also did, independently) that a ball of gas the size of the sun, 108 times bigger in diameter than the Earth, could be kept hot inside if it was shrinking slowly under its own weight. Such shrinking releases gravitational energy, which is converted into heat. The rate at which the sun must be shrinking to maintain its present heat output can easily be calculated, and it is only about fifty metres (150 feet) per year. Unfortunately, slow though that is, it means that the sun would fizzle out within about twenty million years, a number that is known as the Kelvin–Helmholtz timescale. That was still far too short a span for the biologists and geologists, who required timescales of many hundreds of millions of years to produce the changes they saw in the living and non-living world (though today we know that this process is the way young stars get hot inside when they are born).

It turned out that both parts of Thomson's statement are correct. 'Within a finite period of past time' the Earth was indeed unfit for life, but that 'finite period' is much, much longer than the Kelvin–Helmholtz timescale. Further, there are indeed laws of physics that were unknown in the middle of the nineteenth century. In particular, radioactivity.

X-rays were discovered by the German physicist Wilhelm Röntgen in 1895, and this led to the discovery of other forms of radiation and the study of materials that release such radiation – radioactive material. The person who took up these new ideas and used them both to point the way towards a 'new' energy source for the sun and to come up with a new timescale for the Earth was New Zealander Ernest Rutherford, who worked at various times in Cambridge, and Manchester, as well as at McGill University in Montreal. But there is no need for us to go through all the steps that led him to the discovery of an accurate terrestrial timescale. We can skip to the end of the story.

The first thing this kind of radiation provides is energy, produced by mechanisms that Thomson knew nothing about in the 1850s. The radiation is produced when the central parts of the atoms of some heavy elements (their nuclei) re-arrange themselves into states with lower energy. Lower energy states are always preferred in nature. As the nucleus adjusts itself in this way, the excess energy is carried off in the form of the radiation that we detect, and the atom may be converted into an atom of a different substance. For example, one form of uranium 'decays' in this way to make lead.

The energy released in these processes comes ultimately from the conversion of a tiny amount of matter into energy, in line with Albert Einstein's famous equation $E = mc^2$. But Einstein only came up with that equation in 1905, and it took several decades for astronomers to discover that what keeps the sun and stars hot is not radioactive decay but a set of processes that fuse very light nuclei together (in particular, converting hydrogen into helium), which also releases energy. What mattered at the beginning of the twentieth century was that it was clear to people like Rutherford that there were laws of physics that were unknown to Thomson's generation, and that processes going on inside atoms could provide energy for the sun for a very long time indeed. How long? That was where Rutherford's most important contribution to the dating debate came in.

Rutherford discovered that if you start with any particular amount of a radioactive element, half of it will have decayed into another form after a certain time, called the half-life. During the next half-life, half of what is left (a quarter of the original) will decay, and so on. The half-life is different for various radioactive substances, and in each case it can be measured in the laboratory. It means that if you have a sample of rock that contains a mixture of a radioactive element and its so-called 'daughter' products (such as radioactive uranium and its daughter lead) you can measure the proportions of each substance present and use that to work out how old the rock is – how long it is that the radioactive decay has been going on. In 1905, using this technique, Rutherford and his colleague Bertram Boltwood measured the age of a sample of rock as 500 million years – twenty times the Kelvin-Helmholtz timescale. But even this turned out to be a relatively young piece of rock. Since 1905, the age of the oldest known rocks found on Earth has been pushed back, using this utterly reliable and accurate technique, to more than four billion years, neatly matching modern estimates of the age of the sun. The timescale mystery was solved.

In the following chapters, we do not go into great historical detail about how scientists who study the Earth – Earth scientists – have discovered what they know about our home planet. Instead, we concentrate on the most up-to-date version of the story they have uncovered, the story of the Earth as it is today. It makes sense to start this story by setting the Earth in its context as a planet, looking at how the Earth formed as part of the solar system – not least because that explains where the radioactive elements still decaying on Earth today, so important to an understanding of terrestrial time, actually came from.

2

Our place in space

The Earth is made of stardust. In the space between the stars, there are huge clouds of gas and dust, the raw material from which new stars and planets are formed. Most of the gas is primordial – hydrogen and helium left over from the Big Bang, in which the universe as we know it began. But the dust is different. It is material that has already been processed inside previous generations of stars and thrown back out into space to be recycled.

The story of the Earth begins with the collapse of one of those clouds of gas and dust to make the sun and its family of planets – and probably also several other stars and their planetary partners. In order to understand our place in space, and in particular where the radioactive elements that are so useful in measuring the age of rocks came from, we need to understand where those clouds of gas and dust came from – a story that really does begin at the Big Bang itself.

The universe at large

Nobody really knows exactly what happened to make the cosmic fireball that we call the Big Bang. But, as we have previously described in our book *From Here to Infinity*, there is an overwhelming weight of evidence that something triggered the expansion of the universe out of that super-hot, super-dense fireball 13.7 billion years ago. The stuff that expanded away from the Big Bang wasn't just everyday matter, the kind of atoms that we see around us in the modern universe. We can see atomic matter because when it gets hot, it radiates light; that's why stars shine. But there is another kind of matter in the universe, which doesn't shine, that is prosaically called dark matter. (Astronomers can sometimes be very unimaginative.) Dark matter is important, because although it doesn't shine it does exert a gravitational pull, and there is actually more of it – much more of it – than there is atomic matter.

It was the gravitational pull of dark matter that tugged clumps of stuff together as the universe expanded after the Big Bang, stopping everything from getting spread out so thinly that no stars could ever form. Within these concentrations of a mixture of atomic matter and dark matter, enormous clouds of hydrogen and helium gas shrank down and broke up to make islands of stars. Each stellar island, called a galaxy, is held together by gravity, and whole clusters of island galaxies are held together by gravity, with galaxies within a cluster orbiting around one another like bees buzzing around in a swarm. But the clusters of galaxies continue to move apart from one another even today, as the universe as a whole expands.

Everything important for the existence of the sun and the Earth (apart from the Big Bang itself) has taken place within one of those island galaxies, which we call the Milky Way. The Milky Way galaxy is a flat disc of stars, so big that it takes light about 100,000 years to cross from one side of the disc to the other. For this reason, it is said to have a diameter of roughly 100,000 light years. To put that in perspective, light travels at 300,000 kilometres per second, and takes only a little over eight minutes to cross the 150 million kilometres (100 million miles) between the sun and the Earth. In the middle of the disc of our galaxy there is a bulge, rather like the yolk in a huge fried egg, about 23,000 light years across and about 3000 light years thick. Farther out, the disc is only about 1000 light years thick, so it is a hundred times bigger in diameter than its thickness. And the whole system is rotating around the centre of the disc.

Apart from the dark matter, which we cannot see, this island in space contains hundreds of billions of stars, some bigger than the sun, some smaller, but each more or less like the sun. The sun is an ordinary star. But the scale of the Milky Way is so great that even the nearest stellar neighbour to the sun is several light years away. That's why the stars appear only as pinpoints of light on the night sky, not as huge glowing spheres like the sun.

The central bulge of our galaxy contains only old stars, and very little in the way of gas and dust. There, all the material from which stars are made has been used up. But the disc contains a variety of stars of all ages, and great clouds of gas and dust, the stellar nurseries in which new stars are still being born. The sun is about two-thirds of the way out from the centre of the Milky Way to the edge of the disc, and roughly in the middle of the thickness of the disc. It takes about 225 million years to orbit round the galaxy once. But how did it get there?

Making stars

The oldest stars in the Milky Way are about twelve billion years old, so the pieces from which it was assembled were already being put together less than two billion years after the Big Bang. But radioactive dating and other techniques tell us that the sun and our solar system are only a little over 4.5 billion years old. A lot of stars were born, lived, and died between twelve billion years ago and 4.5 billion years ago, and that is why we are here now.

From a combination of an understanding of the basic laws of physics and measurements of the phenomenon known as the cosmic microwave background radiation, a hiss of radio noise left over from the Big Bang itself, cosmologists infer that the atomic material that emerged from the Big Bang was roughly seventy-five per cent hydrogen and twenty-five per cent helium, mixed in with large quantities of dark matter. Hydrogen and helium are the two simplest chemical elements, the basic building blocks from which the other elements are made. This process of element-building began in the first stars, which started out as balls of hydrogen and helium gas, collapsing under their own weight.

When such a ball of gas, containing many times as much matter as there is in our sun, collapses in this way, it gets hot inside because, as Kelvin and Helmholtz realized, gravitational energy is being released. At the same time, the pressure inside builds up, as the ball of gas shrinks. When the pressure and temperature are great enough, the nuclei of hydrogen atoms collide with one another deep inside the ball of gas, so violently that some of them fuse together to make more helium. This process of nuclear fusion releases energy, which provides an outward pressure that stops the collapse of the ball of gas. It has become a star.

Nuclear fusion is the opposite of the process of radioactive decay. Both processes operate naturally, because it happens that the lowest energy arrangement for matter in the form of atomic nuclei occurs for atoms with middling masses. Very light elements, like hydrogen and helium, can fuse to make middle-range elements, with energy being released as a result; but equally, very heavy elements such as uranium and radium can be broken down into lighter elements, with energy also released. In extreme cases, a heavy nucleus splits into two moderately sized pieces; in this extreme form, the radioactivity is called nuclear fission.

Inside the first stars, hydrogen was converted into helium and helium was converted into carbon, oxygen, and other elements. But when this nuclear 'fuel' was used up, the stars suddenly collapsed inward upon themselves, releasing so much gravitational energy so quickly that they blew apart and scattered the elements manufactured inside them through space. So later generations of stars started out not just with hydrogen and helium but with traces of heavier elements as well. Very massive stars run through this life cycle quite quickly, in the span of a few million years, because they need to burn a lot of fuel to hold themselves together. This is the origin of the dust that astronomers see in interstellar clouds. There was plenty of time for these processes to operate over several generations of stars before the sun formed from a collapsing cloud of gas and dust within the Milky Way galaxy. The presence of heavier elements than hydrogen and helium in these clouds helped smaller stars, like the sun, to form in later generations, because these elements form compounds that are good at radiating heat away into space. This means the clouds can shrink more, and break up into smaller pieces, before they form stars.

But many large stars still formed in the time between the Big Bang and the birth of our sun, and many still form today. This is important for two reasons. First, the steady process of building up heavier elements by nuclear fusion stops with iron and nickel, which are the most desirable nuclei in energy terms. All the heavier elements, including things like gold and lead, as well as the radioactive elements, were made in the great explosions that occurred when large stars died. In these explosions, called super-novas, nuclei were squeezed together so hard that they fused, even though energy had to be put in to make this happen. The radioactive elements present at the birth of the sun and our solar system, and still used in radioactive dating, were made in this way, inside one or more supernovas that exploded just before the sun was born.

This highlights the second important feature of supernovas: the clouds of gas and dust between the stars only begin to col-lapse to make stars if something gives them a squeeze. Although there are other ways to trigger the collapse, the best way is for a supernova explosion to send blast waves rippling through the nearby clouds. This is what triggered the birth of our solar system.

Making planets

The sun and its family of planets – the solar system – were born roughly 4.5 billion years ago, more than nine billion years after the birth of the universe. The Earth, as part of our solar system, is almost exactly one-third of the age of the universe itself. In round numbers, between ten and twenty new stars are born in our galaxy each year, and the same number die. In nine billion years, roughly a hundred billion times as much matter as there is in our sun has been processed inside stars and ejected into space to be recycled, enriching the clouds of interstellar material so that succeeding generations of stars start out with more heavy elements. But only a small proportion of the hydrogen and helium gets processed in each star, or in each generation of stars. Even though generations of stars had formed, gone through their life cycles, and scattered material into space to enrich the interstellar clouds in the nine billion years before our solar system formed, the cloud from which the solar system formed still contained only a smattering of elements heavier than helium.

The mass of the solar system reflects these cosmic origins. Of the total, based on a variety of astronomical evidence, hydrogen makes up 70.1%, helium 27.9%, and oxygen 0.9% of the total mass of the whole system, the sun as well as the planets. Those figures leave just 1.1% for everything else combined. The other elements are present in such small amounts that it is easier to rank them in terms of numbers of atoms, rather than by mass. Taking the top ten (apart from hydrogen and helium), we know that for every seventy atoms of oxygen there are forty atoms of carbon, nine atoms of nitrogen, five atoms of silicon, four each of magnesium and neon, three atoms of iron, and just two of sulphur. Other elements are present in even smaller quantities; there are only three atoms of gold for every ten *million* atoms of sulphur in the solar system. Even on Earth, gold is rare – that's why we value it so highly. But, apart from hydrogen, all the things we think of as common elements are actually quite rare, and only seem to be common to us because we live on a ball of rocky material where these elements are concentrated. When the cloud of gas and dust that formed the solar system collapsed, most of the hydrogen and helium moved into the centre, where they formed the sun, but some of the heavier material was left behind in a disc swirling around this young sun, and it was in that disc that particles of dust stuck together to form the Earth.

The cloud from which the solar system formed could not collapse completely to form a single star, with no planetary companions, because it was rotating. When a spinning object shrinks, it spins faster as a result. You have probably seen this when watching a spinning ice-skater pull his or her arms inward. The natural tendency, as anyone who has tried spinning on ice skates will know, is to fling your arms outward. For a collapsing cloud of interstellar gas, this process in effect produces a force that counteracts the inward pull of gravity, and beyond a certain point the proto-star is only able to continue collapsing by shedding material into a disc orbiting around the young star. The sun still rotates, once every twenty-seven days, and all the planets and bits of material left over from the formation of the solar system orbit around it in the same direction that the sun rotates, as you would expect.

Several different kinds of object formed in the disc of material around the young sun. We are especially interested in one of those objects, the planet Earth; but it is worth putting it in its proper perspective as one component of the solar system. Closest to the sun, there are four rocky planets – in order of their distance from the sun, they are Mercury, Venus, Earth, and Mars. Then comes a belt of cosmic rubble, pieces of rock left over from the formation of the planets; they are known as asteroids, and the region they inhabit is called the asteroid belt. Beyond the asteroid belt there are four large planets mostly made up of gas – Jupiter,

Saturn, Uranus, and Neptune. The largest of these, Jupiter, contains as much mass as 317 planets like the Earth, but it is mostly hydrogen and helium. Clearly, the 'gas giants' formed in a different way from the Earth, but we won't be concerned with them here. Beyond the gas giants there is another belt of debris, this time containing large chunks of icy material. It is known as the Kuiper Belt; Pluto was once thought to be a planet in its own right, but is now known to be a Kuiper Belt Object (and it isn't even the biggest one of those). Much farther out in the depths of space, literally halfway to the nearest star, the whole solar system is surrounded by a spherical cloud of icy objects, typically tens of kilometres across, called comets. Occasionally, one or more of these ice mountains is disturbed, perhaps by the gravity of a nearby star, and falls in towards the sun, passing through the inner solar system in a blaze of glory before zipping around the sun and back out into space. But for now we are more interested in what happened much closer to the sun 4.5 billion years ago, when this hot young star began to radiate energy through the disc of material that surrounded it.

Making the Earth

At first, the disc must have contained the same mixture of stuff as the original cloud from which the solar system formed – ninety-eight per cent hydrogen and helium and just two per cent everything else. Because helium doesn't combine chemically with anything, all of the helium was in the form of gas. Hydrogen does combine with other elements, to make things like water, methane, and ammonia; but there wasn't very much stuff around for it to combine with, so most of the hydrogen was also in the form of gas. The atoms of the other substances – the oxygen, carbon, nitrogen, and so on – combined with one another and with hydrogen to make molecules, and many of the molecules stuck to the surfaces of tiny grains of graphite (like little particles of soot, as small as the particles in cigarette smoke). As the sun began to heat up, its radiation drove all the remaining gas out into space. At least in the inner part of the solar system, only the solid particles remained in the disc around the sun, all moving in the same direction, but at different speeds and in slightly different orbits, crossing each other's paths so that they kept bumping into one another. Because they were moving in the same direction these collisions were gentle enough for particles to begin to stick together, and electric forces probably helped the build-up of fluffy dust balls not unlike the ones you find under the bed if you neglect the housework. These fluffy dust balls then began to pull on one another by gravity, clumping together in larger and larger lumps. It only took about a hundred thousand years for these lumps to grow to about a kilometre or so across, becoming so-called planetesimals, the building blocks from which the rocky planets formed. Some of the objects in the asteroid belt may be primordial planetesimals left over from the formation of the solar system; others are pieces of rock that were once part of larger objects that have been broken apart in collisions, because from the time of the planetesimals onward planet-building was a very violent process, involving collisions between larger and larger lumps of rock.

Computer simulations show that by about a million years after the collapse of the cloud from which the sun and planets formed, there would have been twenty or thirty objects in the region between the sun and the present orbit of Mars, ranging in size from about the size of the moon (roughly twenty-seven per cent of the diameter of the Earth today) to about the size of Mars (roughly fifty-three per cent of the diameter of the Earth today). They would have been accompanied by a huge number of smaller planetesimals, which would have been swept up by the larger objects in a series of collisions, while the larger objects themselves collided with one another and merged until eventually only four or five large objects were left – the objects that became Mercury, Venus, Earth, and Mars, plus at least one other Mars-sized object. Seared by the heat of the young sun, which would have destroyed complex molecules and driven gases outward, these four (plus one) proto-planets would have been mostly made of iron and silicates, plus stable compounds of carbon.

Earth and Venus are very nearly twin planets in terms of size, and occupy adjacent orbits around the sun. They must have formed in much the same fashion as each other. The key development in this later stage of planet-building, apart from the fact that the planets got bigger, was that the impacts generated a lot of heat, produced from the kinetic energy of the incoming objects. This heat melted the surface of the young planet, and only slowly penetrated into the interior, but after about fifty million years the young planet would have been fluid enough for the iron and any other heavy elements to sink downwards into the core, while the lighter silicates floated on top, forming a crust as the planet began to cool. Along with the iron, other heavy elements also sank into the core, including radioactive uranium produced in the super-nova explosion that triggered the birth of the solar system. Heat released by the radioactive decay of this uranium over billions of years helps to explain why the core is still molten today, and why Kelvin's estimate of the age of the Earth was wrong; but this is not the only factor involved, since Kelvin also forgot to allow for the effect of convection in the liquid core in his calculations, and this also changes the age estimate. (We'll hear more about convection on our journey, especially in chapter 5, when we look inside the Earth.) Furthermore, once the solid crust was in place, it acted like an insulating blanket around the molten core, slowing the rate at which heat could escape into space.

Curiously, although this process perfectly explains the nature of the planet Venus today, it cannot completely account for the present structure of the Earth. The crust of the present Earth is very thin, averaging a little more than five kilometres, or three miles, under the oceans, which account for two-thirds of the planet's surface, and thirty kilometres, or eighteen miles, under the continents. As we shall explain, this is very important for the processes, known as plate tectonics, which have changed the geography of our planet over the course of geological time. It is easy for the Earth's thin crust to crack, like an eggshell, releasing internal heat from volcanoes and making the pieces jostle about on the surface, causing earthquakes. But data from space probes suggest that the crust of Venus is at least fifty kilometres – thirty miles – and perhaps as much as one hundred kilometres thick, so that it forms an immovable layer with no plate tectonic activity. As heat builds up in the core as a result of radioactivity, the pressure increases until, at very long intervals (most recently about 600 million years ago), the crust finally cracks apart in a global upheaval, with huge floods of molten lava spewing out, covering Venus's entire surface and solidifying into a new surface layer as the heat is released. Then the planet settles down into an inactive state for another long interval, many hundreds of millions of years.

The presence of a thick, silicate-rich crust on Venus is very much in line with what astronomers would expect from their understanding of the composition of the disc of material from which the inner planets formed. There is no puzzle about why Venus should have such a thick crust. The puzzle is why the Earth should have such a thin one – why we are missing a lot of silicates that ought to be here. And the answer seems to be intimately associated with the presence of another unique feature of the Earth – its large moon.

Making the moon

Although it is only a quarter the size of the Earth (in terms of its diameter), that still makes our moon by far the biggest of any of the moons in the solar system, in comparison to the size of the planet it orbits around. To many astronomers, it should not be regarded as a moon at all, in the sense that the moons of Jupiter and the other giant planets are moons; they prefer to describe the Earth-moon system as a 'double planet'. In fact, none of the other three inner planets really has a moon at all. Mercury (the diameter of which is about thirty-eight per cent of that of the Earth) and Venus orbit the sun in splendid isolation, while Mars is accompanied only by two tiny 'moons', Phobos and Deimos, each only a few kilometres across. They are clearly stray lumps of rock that have been captured by the Martian gravity field from the nearby asteroid belt. This makes the Earth-moon double planet even more unusual. So how did it form?

The most likely explanation is that the Earth did indeed begin as a near-identical twin to Venus, complete with a thick crust, while another planetary object, about the size of Mars, formed nearby. The most likely place for this object to form would have been at one of two places known as Lagrangian points, sixty degrees ahead of or behind the Earth but in the same orbit around the sun. At these points the combined effect of the gravitational pull of the sun and the gravitational pull of the Earth produces a kind of gravitational pothole, a place where small objects can accumulate and stay. The equivalent Lagrangian points are used today as stable parking places for satellites, such as infrared telescopes, which need to be kept far enough away from the Earth not to suffer interference from our home planet. A small object which is not quite at the exact Lagrangian point wobbles slightly to and fro about the point itself, like a swinging pendulum; but if a large object grew up out of cosmic debris near to one of the Lagrangian points of the Earth's orbit, these oscillations would get bigger and bigger, soon becoming so extreme that the object would bash into the Earth itself. Such an event would have happened within ten *million* years of the formation of the original crust of the Earth.

Don't imagine that this event would be like two pieces of solid rock colliding and chipping pieces from one another. One of the names given to such a model is the Big Splash, and the image it conjures up accurately indicates what happened when the Earth was young. So much kinetic energy would have been released by the glancing collision between a Mars-sized object and the young Earth that the incoming object would have been completely destroyed, and the entire surface of the Earth itself would have remelted. The dense, metallic core of the incoming object would have sunk through this molten outer layer and been absorbed into the core of the young Earth, while lighter material from the incoming object and from the Earth's original surface would have been splattered out into space. About ten times the present mass of the moon would have been ejected in this way. Most of it escaped entirely into independent orbits around the sun, but some was captured in a ring of material around the Earth. As the surface of the Earth cooled once again and formed a new, thinner crust, the material in this ring coalesced into the moon, repeating in miniature the process by which the planets themselves formed around the sun.

There is a wealth of evidence in support of this model of how the Earth-moon system formed. The most compelling is that samples brought back from the moon show that it has exactly the same composition as the Earth's crust, while seismic measurements of lunar quakes show that it has no large metallic core. The age of lunar rock also tells us when this dramatic event happened – 4.4 billion years ago, almost as soon as the formation of the sun. Such a glancing blow also explains why the Earth rotates so rapidly, once every twenty-four hours, while moonless Venus rotates only once every 243 of our days. Just after it was struck the glancing blow that formed the moon, the Earth would have been spinning much faster, and it has been slowing down ever since.

The off-centre impact also gave the Earth its tilt, which is the reason why we have seasons, but the presence of such a large moon orbiting the Earth has since acted as a gravitational stabilizer, stopping the tilt from varying very much over geological time. Incidentally, a combination of the extra iron in the Earth's core and the rapid spin probably explains why our planet has a strong magnetic field.

And there is one more piece of circumstantial evidence that collisions like this did happen, from farther away across the solar system. The strength of the gravitational pull of the planet Mercury shows that, in spite of its small size, it has a relatively large mass, meaning that it has a high density. While our moon resembles the crust of the Earth without a core, Mercury resembles the core of the Earth without a crust. The natural explanation is that a much larger object originally formed in the orbit of Mercury, but that early in the life of the solar system it was hit, not in a glancing blow but a head-on collision, by another proto-planet. In a head-on collision, all the lighter material would have been blasted away into space, leaving only the heavy core behind.

But that isn't all that Mercury and the moon can tell us about how the Earth got to be the way it is. All of this would have happened only a little more than 4.5 billion years ago, in a burst of activity that was almost indecently rapid by astronomical standards. The record of what happened next, to complete the formation of the Earth as a planet that would become a suitable home for life, is revealed by the battered faces of those two objects.

Making ocean and atmosphere

When the Earth cooled again after the impact in which the moon was formed, it was dry and lifeless, and still very hot even after the crust solidified. Some gases would have escaped from the Earth's interior through volcanic activity, as a result of chemical reactions going on below the surface in magma – a mixture of molten rock, volatile material, and solid pieces of rock that is found beneath the surface of the Earth. But the bulk of the Earth's atmosphere, and almost all the water in the seas, were brought to Earth from space after the formation of the Earth-moon system.

We know when this happened because astronauts have visited the moon and brought back samples of rock, which have been dated. All the evidence shows that about 100 million years after the solar system formed, just after the moon itself coalesced out of the ring around the Earth, it was covered by an ocean of magma some 200 kilometres (125 miles) deep. This set in a smooth surface, which was soon intensely cratered by a massive bombardment from space to produce the surface we see today. This bombardment lasted for, in round terms, half a billion years, ending about four billion years ago, producing craters overlapping one another where later impacts have struck on top of existing craters. For the past four billion years or so, there has been a much smaller cratering rate linked to the continuing presence of bits of cosmic debris between the planets.

If the moon was the only object in the inner solar system with this kind of battered surface, we might guess that it was a result of sweeping up the rest of the debris from the collision that caused the formation of the Earth-moon system. But the surface of airless Mercury shows exactly the same kind of battering record. Although Mars has an atmosphere that has eroded and softened the outlines of craters, there is discernible evidence that the red planet also suffered this kind of bombardment. It is clear that the bombardment affected everything between Mercury and Mars – all the planets of the inner solar system. Among other things, this is how geologists claim as fact that Venus was resurfaced some 600 million years ago; there is no evidence for the bombardment there, just a relatively small number of craters that can be explained by the impact of stray pieces of debris over the past 600 million years.

The Earth's surface has been reworked over geological time, and has also experienced erosion, but the Earth, too, must have been intensely bombarded by debris from space between 4.5 billion and four billion years ago. What kind of debris would this have been? Some pieces were certainly lumps of rock like the present-day asteroids. But, equally certainly, others were comets, lumps of icy material, containing water ice but also things like frozen methane and ammonia, that had formed farther out in the solar system in the cold depths of space and were flung inwards by gravitational interactions with Jupiter and the other giant planets. In terms of the energy released by the impact, it doesn't matter whether an object striking the Earth is made of rock, ice, or a mixture of the two; all that matters are its mass and speed. An object ten kilometres (six miles) across made of an ice/rock mixture arriving at fifty kilometres (thirty miles) per second – about the size and speed of the object that caused the 'death of the dinosaurs' some sixty-five million years ago – would release as much energy as the explosion of 100 million megatonnes of TNT, cratering the surface and spewing its icy burden out as gases, including water vapour that would rain onto the surface. But the moon, with its much smaller mass and therefore much weaker gravitational pull than the Earth, had no hope of retaining these gases, which it lost to space, leaving it as dry and lifeless today as it was when it formed.

The amount of cratering on the moon suggests that between 4.5 billion and four billion years ago the Earth was hit by a total of ten billion billion tonnes of material from space. If only half of this was in the form of comets, and only twenty per cent of the mass of each comet was water ice, that was enough to provide eighty per cent of all the water on Earth today, allowing for some to have been produced by volcanic outgassing. The primordial atmosphere of the Earth was provided as a bonus. And almost immediately, certainly by 3.8 billion years ago, there was life on Earth. The speed with which life got a foothold on our home planet was itself, almost certainly, as a result of the particular mix of material brought down to the surface by comets during this bombardment.

But the story of life will have to wait, for now, while we concentrate on just what it is that makes a planet with a hot, dense interior and a thin crust, mostly covered in water, so much more active and interesting than its neighbours.

3

Drifting continents and spreading seas

Life on Earth is confined to a narrow zone, like a tight skin. In round terms, this extends from the bottom of the deepest ocean, eleven kilometres below sea level, to the top of the tallest mountain, nine kilometres above sea level. That makes the life zone of the Earth only twenty kilometres thick, but that zone contains everything that makes up our everyday world. Compared with the size of the Earth, it is only about as thick as the skin of an apple compared with the size of the apple itself, a very tight skin indeed.

This life zone is at the mercy of powerful forces at work in the crust of the Earth, the seemingly solid rocks beneath our feet. As we shall see, those forces are a result of deep-seated activity far below the surface of our home planet. But before delving into the depths, we should look at how that seemingly solid surface is actually constantly being re-arranged into new patterns by processes that both split continents apart and send them crunching into one another, opening and closing ocean basins and forcing up new mountain ranges as they do so.

A changing world

For those with eyes to see, there has always been evidence that the surface of the Earth has been subjected to dramatic changes. The most obvious example of this is the existence of fossil remains of sea creatures in rocks high above today's sea level. Aristotle puzzled over this back in the fifth century BC, while in the fifteenth century Leonardo da Vinci noted, 'above the plains of Italy where flocks of birds are flying today, fishes were once moving in large shoals'. The natural explanation to most people in Leonardo's time was that these remains were evidence of the biblical Flood; but not everyone was persuaded by that argument. In 1665, the natural philosopher Robert Hooke, carefully hedging his bets without offending the Church, wrote in *Micrographia* that 'some Deluge, Inundation, Earthquake or some other means' had lifted what was once sea floor above sea level. By then, geographers had another puzzle to contemplate.

By the beginning of the seventeenth century, European voyages of exploration had produced such good maps of the continents that in 1620 Francis Bacon, the so-called father of empiricism, commented on the similarity of the coastlines of eastern South America and western Africa, mirror images that look as if they ought to fit together like pieces in a jigsaw puzzle. In his book *Novum Organum*, Bacon said that this is 'no mere accidental occurrence', although he couldn't explain how or why. Once again, the obvious explanation to many people was that the Atlantic Ocean was in actuality a huge river valley, carved out by the Flood across what was once a single landmass.

This kind of thinking continued to colour the first scientific attempts to explain the growth of mountain ranges and the shape of ocean basins. These revolved around the idea that as the Earth had cooled down after its formation, it had shrunk, like a dried-out apple, and the tight skin had wrinkled. That 'explained' mountain ranges. The ocean basins were 'explained' as a result of pieces of the shrinking crust collapsing into cavities beneath the surface – cavities originally full of water, which burst out and produced the Flood. A rival idea suggested that the entire Earth had originally been covered in water, which had steadily drained away into hollows beneath the surface.

But the best early hypothesis about our changing world came from the polymath Benjamin Franklin, in 1782. He suggested that the solid surface of the Earth must be a relatively thin crust, like a shell, floating on a fluid interior. 'Thus, the surface of the globe would be capable of being broken and disordered by the violent movements of the fluids on which it rested'. That, in a nutshell, is the modern theory of plate tectonics, although now we have a wealth of evidence to support Franklin's theory.

And yet, it moves

The first scientist to propose that the continents had moved across the surface of the Earth was the splendidly named Antonio Snider-Pellegrini, an Italian-American living in Paris in the 1850s. His ideas, expressed in his book *La Création et ses mystères dévoilés*, published in 1858, were a mish-mash of biblical interpretation and scientific hypothesizing. They included the suggestion that all of the Earth's landmasses had originally been united in a single supercontinent, but that in a catastrophe associated with the Flood, South America had split off from what is now Africa and been sent hurtling around the globe to its present position. This, of course, flew in the face of the uniformitarian ideas that were already well established then, twenty-five years after the publication of Lyell's great book. But a few people took on board Snider-Pellegrini's idea that South America and Africa had split apart from a single landmass, while quietly ignoring his suggestion that this happened all at once, in a sudden catastrophe.

The evidence that South America and Africa were once joined is now incontrovertible, and was already becoming so in the nineteenth century. Mountain ranges in South America that run into the Atlantic Ocean match up with mountain ranges in Africa that run into the Atlantic Ocean, and the strata within those rocks match up, too; low-lying regions match up across the join, as do fossil beds; and the kinds of living plants and animals seen on opposite sides of the Atlantic match up (although because plants and animals evolve, this match is less perfect). Overall, the effect is as if you cut the shapes of Africa and South America out of a single sheet of newspaper and pulled them apart. You would easily be able to join the two pieces back together again by matching up the words and sentences from the paper's stories. By the end of the century, the question that really needed to be answered was not *whether* the continents had moved across the surface of the Earth over geological time, but *how* they had moved.

It would be a long time before science could answer that question, but a major clue became available in 1908. That year, the US geologist Frank Taylor presented his ideas about the drifting continents to a meeting of the Geological Society of America. He saw that great mountain ranges were actually the result of the Earth's crust being squeezed sideways and were distorted as a result, a finding in line with pioneering studies carried out by Bailey Willis, of the US Geological Survey and Stanford University. (Ironically, Willis himself believed that 'the great ocean basins are permanent features of the Earth's surface'.) These sideways forces, Taylor argued, were associated with a 'mighty creeping movement' of the continents across the face of the Earth. In his most prescient insight, he pointed to the existence of a great undersea mountain range, known as the Mid-Atlantic Ridge, that runs down the sea floor halfway between South America and Africa; it also extends into the North Atlantic. This, he said, is the site of the original rift between the continents, suggesting that the ridge 'has remained unmoved while the two continents on opposite sides of it have crept away in nearly parallel and opposite directions'. Nobody took much notice of Taylor's thesis, and he didn't really try to promote it. But within a few years, a German meteorologist, Alfred Wegener, picked up the idea of drifting continents and made it his own, promoting it vigorously in a series of books and making it impossible to ignore.

Although he wasn't a geologist, in 1911 Wegener came across a scientific paper devoted to the proposal that the similarities between strata, fossils, and living things on opposite sides of the ocean could be explained if there had been a land bridge between them that had long since sunk beneath the waves. He found the concept ridiculous, not least since, by the early years of the twentieth century, it was known that the rocks of the sea floor are denser than the rocks of the continental crust. Continents are like icebergs floating in the denser material beneath the crust. If some mysterious force had made continental land bridges sink into the underlying material, they should have bobbed up again, just as an ice cube pushed below the surface of a glass of water will bob back up. And Wegener knew a thing or two about the behaviour of floating ice, because he had worked as the official meteorologist on a Danish expedition to Greenland. Floating ice cannot sink beneath the waves – but floating ice sheets can break up and drift apart, with the watery gaps between them widening. By January 1912 Wegener had developed these notions into the first version of a theory of continental drift, which he presented to a meeting of the German Geological Association in Frankfurt.

Wegener made very few converts, and his work developing the theory was slowed by the time he was forced to spend on his proper job as a meteorologist and university lecturer; it was slowed still more by his compulsory service in the German army during World War I. The big snag with his idea was that it involved the continents drifting through the seemingly solid rock of the ocean floor, like icebergs drifting through the ocean. The mechanism for the motion remained mysterious, and that was enough to put most geologists off the idea. But over the years Wegener assembled a growing weight of evidence against the land bridge hypothesis. Detailed studies of the rock strata and other evidence showed that the 'lines of newsprint' line up precisely across the oceanic gap, with no 'missing words' sinking out of sight. (It was Wegener, by the way, who first came up with the newspaper analogy.) He also used his knowledge of meteorology to reconstruct past climates from studies of the different kinds of rock and fossils found in different strata with different ages. This analysis demonstrated, for example, that the island of Spitsbergen, now in the frozen north in the Arctic circle, had once been covered by tropical plants.

In his Sherlock Holmes stories, Arthur Conan Doyle had his great detective point out that once you have eliminated the impossible, whatever is left, however improbable, must be the truth. Wegener proved that the land bridge was not possible; thus, the continents *must* drift, however unlikely that seemed. His reconstructions of the 'lines of newsprint' showed that all the continents had once been joined in a single supercontinent, which he called Pangaea, meaning 'whole Earth'. This had started to split apart about 300 million years ago, according to Wegener's calculations. In one of his most important insights, Wegener also provided an explanation for mountain-building that did not involve a shrinking Earth. He suggested that mountain ranges are formed at the leading edge of a drifting continent, as the rocks crumple up because of the resistance to their motion. This could explain the existence of ranges such as the Rocky Mountains on the western half of North America. Mountains can also be formed when the rocks crumple because of a collision between landmasses, notably where India, moving north, has ploughed into Asia and forced up the Himalayan range.

In spite of all this, Wegener persuaded very few geologists that the idea of continental drift was right. In November 1930, shortly after his fiftieth birthday, he died, probably of a heart attack, while on his latest expedition to Greenland. His obituaries scarcely mentioned his theory.

To be fair to his contemporaries, the lack of a mechanism for continental drift really did seem like an insurmountable obstacle, even in the 1930s. The image of continents ploughing through the ocean floor like icebergs ploughing through the sea stuck in the throat of geologists, who knew that there was no way for the rocks of the sea floor to part and make way for the continents in the fashion that sea water could part and make way for icebergs. But the solution to the puzzle was already there, in Wegener's first insights into continental drift, had he, and they, but realized it. The right image is not of icebergs moving *through* water, but of sea ice floating *on* water. When ice floes break up and drift apart, they don't do so because they are moving through the water, but because they are being carried along for the ride by ocean currents. It turns out that continents are also being carried along for the ride by underlying currents – but it took a new technology to make this clear.

Into the deep

That new technology was developed as a result of World War II, and the Cold War that followed it. Although attempts to map parts of the sea floor close to continents had been made in the 1930s, the technology to probe the deep oceans developed dramatically with sonar. This method was used to detect enemy submarines, measure the depth of water, and search for underwater obstacles near invasion landing beaches. Even the simplest of sonar devices – an echo sounder that sends a pulse downwards to bounce off the sea floor – dramatically changed the image geologists had of the ocean floor. By timing the return of the pulses to the ship as it moved along a series of well-charted tracks, scientists could gain a profile of the sea floor accurate to within a metre or so. Seismic surveys, in which explosions are set off in the sea or on the sea bed and the vibrations from them recorded, provided even more information. Both techniques were improved dramatically and applied around the world during the 1950s through research funded by the US Navy, which wanted detailed maps of the sea floor to help in the search for potentially hostile nuclear submarines – and for good hiding places for America's own vessels.

Before these sonar studies were carried out, geologists had expected the ocean floors to be relatively smooth and covered with a deep layer of sediment, perhaps five kilometres, or three miles, thick, washed down from the continents over the eons. The continents themselves don't actually end at the water's edge, but extend a few hundred kilometres out from the coast, forming a shelf where the water is relatively shallow, no more than 200 metres (650 feet) deep. Ocean basins begin at the edge of the continental shelf, which slopes down to the deep ocean floor; this is, on average, a bit less than four kilometres (2.5 miles) below sea level. Today, twenty-nine per cent of the surface of our planet is land, and seventy-one per cent is covered by sea. If the sea level fell by about 180 metres (600 feet), so that the edge of the ocean matched the edge of the continental shelf, the proportions would be thirty-five per cent land and sixty-five per cent ocean. This is a true indication of the balance between continental crust and sea floor crust.

The new surveys revealed that the sea floor is very rugged, with hills and valleys covered by only a relatively thin layer of sediment. There is the pronounced Mid-Atlantic Ridge, an underwater mountain chain running down the middle of the Atlantic Ocean that links up with similar mountain chains to form a network around the globe. The crust itself is only about seven kilometres (four miles) thick beneath the oceans, compared with a range from about thirty-five to seventy kilometres (twenty to forty miles) for the continental crust. The sediment layer, even on the floor of the Pacific Ocean, the largest ocean in the world, turned out to be only a few hundred metres thick, corresponding to the amount of sediment that would be washed into it over 200 million years. Radioactive dating has confirmed that there are no ocean rocks older than about 150 million years. This means that the Pacific Ocean can only be 150 million to 200 million years old.

More recent investigations have added to our new picture of the sea floor. Increasingly detailed echo-sounding images show that there is a deep canyon in the middle of the Mid-Atlantic Ridge, like a crack in the Earth. This turns out to lie exactly along the line of most undersea earthquakes in the Atlantic Ocean, confirming that something is happening at the ridge. Similar earthquakes occur along the entire global network of ocean ridges, which accounts for about one-twentieth of all the seismic energy released around the world. Some of the ridges were even discovered by earthquake monitoring surveys before they were mapped by echo-sounding or other techniques – a striking correlation.

Probes lowered to the ocean floor reveal that the rate at which heat is flowing out of the region of the Atlantic crack is eight times greater than the rate at which heat is escaping from the rest of the sea floor. The explanation: there is indeed a crack in the Earth's crust centred on the ridge, where heat from inside the Earth is welling up to the surface, pushing up the mountains that comprise the ridge as it does so. This has been confirmed by samples of rock laboriously dredged from the sea bed that are made of newly formed lavas. Further, the shape of the Mid-Atlantic Ridge exactly follows the contours of the east coast of South America and the west coast of Africa. With this evidence in hand, it is now clear that the ridge is the site of the crack that sundered South America from Africa, and that the two continents have been moving apart from one another symmetrically ever since.

But how are the continents actually moving on either side of the crack? One suggestion was that the Earth might be expanding, but that was quickly ruled out by further evidence, and a new insight.

The evidence was actually rather old by the middle of the 1950s, having been discovered by a Princeton geologist, Harry Hess, while he was in command of a transport ship in the Pacific during World War II. The ship had been fitted with one of the latest echo sounders, and Hess kept it switched on all the time, taking profiles across the sea floor wherever the ship went. He was perplexed to discover that the floor of the ocean is dotted with flat-topped underwater mountains, which he called guyots, in honour of the nineteenth-century geologist Arnold Guyot. These mountains resembled the flat-topped islands seen where extinct volcanoes have poked above the waves and been scoured flat by the action of wind and waves. They often form coral islands, where a ring of coral grows up around the edge of the eroded mountain, with a shallow lagoon inside.

What Hess noticed was an aspect of a mystery that had baffled Charles Darwin on his famous voyage on the *Beagle* in the 1830s. The puzzle is that coral only grows in shallow water, where there is enough sunlight, and needs a solid base to stand on. It's easy to see how a reef can grow up in the shallow water near an island. But reefs are also found far out from these small islands' shores, where living coral is standing on a base of dead coral which is itself resting on the sea floor; some reefs actually enclose lagoons with no island inside at all. Darwin's insight was that the islands that had allowed such reefs to get started had sunk beneath the waves, and that as the reefs submerged new living coral continued to grow on top of the old, dead reef. The coral does not, as some people had thought, grow around the rim of the crater of an old volcano, since by the time the island has sunk beneath the waves any such features have long since eroded away. It's worth seeing what Darwin himself said, because he always expressed his ideas so clearly that it is impossible to improve on his explanations. He wrote:

Let us then take an island surrounded by fringing-reefs, which offer no difficulty in their structure; and let this island with its reef ... slowly subside. Now as the island sinks down, either a few feet at a time or quite insensibly, we may safely infer from what is known of the conditions favourable to the growth of coral, that the living masses, bathed by the surf on the margin of the reef, will soon regain the surface. The water however, will encroach little by little on the shore, the island becoming lower and smaller, and the space between the inner edge of the reef and the beach proportionally broader ... We can now at once see why encircling barrier-reefs stand so far from the shores which they front.

Continue the process even longer, and you will be left with a coral-fringed lagoon without any island at all. This is known as an atoll, a term which Darwin popularized in his book on coral reefs, published in 1842. Over a still longer period, the coral will no longer be able to support itself, and everything will disappear beneath the waves.

Of course, Darwin had no idea why the island should be sinking, and he did not know about the underwater mountains discovered by Hess. But his insight does explain how guyots could erode, even though they now lie far below the surface, and why, as it later turned out, some of them are still fringed with dead coral.

When later discoveries showed that the farther away they were from the ocean ridges, the deeper the guyots lay beneath the surface, Hess grasped what was going on. The idea that hit him, more than ten years after his wartime discovery, was that the guyots had formed near the ocean ridges, poking above the sea surface and being eroded in the usual way, but had then been carried away down into the depths as the ocean floor itself moved, like a conveyor belt, spreading out from the ridge. This was the origin of the concept of sea floor spreading. The continents do not move through the rock of the sea floor, Hess realized, but are carried along on the backs of the 'conveyor belts'. Spreading at a rate of just over a centimetre per year on each side of the ridge (about the rate at which your fingernails grow) would be enough to account for the growth of all of the ocean basins of the world in less than 200 million years, roughly one-twentieth of the age of the Earth.

Hess put forward these ideas in 1960, also pointing out that as new sea floor is constantly being manufactured at ridges, and there is no sea floor older than 200 million years, that means old sea floor must be being destroyed somewhere else to maintain the balance. In his own words:

> The ocean basins are impermanent features, and the continents are permanent, although they may be torn apart or welded together and their margins deformed. The continents are carried passively on the mantle with convection and do not plow through oceanic crust.

This package of hypotheses attracted attention, but in 1960 it wasn't yet enough to persuade most geologists that the continents had really drifted. Within a few years, though, the clinching evidence came in – and, once again, it was partly thanks to the Cold War.

Magnetic recordings

The key to the breakthrough was magnetism. Many rocks contain minerals that are rich in iron, and when igneous rock that contains magnetic material is setting, these iron compounds line up with the Earth's magnetic field, giving the rock a weak but permanent magnetism. This is something like a fossil magnetic field, recording the direction of the Earth's magnetic field at the time and place the rocks were laid down. This alignment doesn't just correspond to the magnetic north–south direction horizontally, but also follows the dip of the Earth's magnetic field downwards, which is different at different latitudes – greater at high latitudes, smaller at low ones. The angle of the dip, called the inclination, corresponds to the latitude where the rock is laid down. When magnetometers sensitive enough to measure these traces of rock magnetism were developed in the 1950s, they showed that the magnetic inclination in rocks from Britain is much less than the value corresponding to Britain's present position on the globe. It looked as if Britain had been moving northward for the past 200 million years. Similar studies in the southern hemisphere suggested that the southern continents (except Antarctica) had also been drifting northward. But this evidence was treated very cautiously at first, because the techniques involved were new and difficult. It was the application of magnetic studies to the sea floor that changed many minds.

It happened almost by chance. In 1955 the US Navy funded a very accurate survey of part of the sea floor off the west coast of North America. This involved a ship sailing up and down a series of parallel tracks while making echo soundings. The scientists involved were told that while they were carrying out this tedious but important task they could also carry out any other experiment that didn't interfere with the survey itself. The only thing they could think of that would fit the bill was to carry out a magnetic survey, using a magnetometer in a kind of torpedo towed behind the ship. The instrument actually measured the Earth's magnetic field, extremely accurately; in places where there was rock magnetism aligned with the Earth's magnetism the measured field was a little stronger, while in places where the rock magnetism was aligned the opposite way the measured field was a little weaker. The researchers weren't looking for anything in particular, just anomalies in the magnetic field that might indicate something of interest on the sea floor.

What they found was a series of magnetic stripes, more or less parallel to one another. Along one stripe, the rock magnetism was lined up in one direction. In the next stripe, the rock magnetism was reversed. Then in the next stripe, it was lined up with the first stripe again, and so on. The stripes also ran parallel to the nearby ocean ridge. The results were published in 1958, stimulating more magnetic studies of the sea floor, but nobody yet knew what they meant.

The answer came from studying rocks on land. As well as the subtle variations in inclination that show, among other things, that Britain has drifted north over the past couple of hundred million years, there are places on Earth where the magnetism in some of the rocks is exactly the opposite of what it ought to be if those rocks had been laid down in the Earth's present magnetic field. The most natural explanation is that the strata with reversed magnetism were laid down when the Earth's magnetism was reversed, with the magnetic poles having changed places. This is plausible because the Earth's magnetism is generated by swirling currents of liquid iron in its core, so changes in the currents can change the magnetic field without the Earth physically toppling over in space. This idea was confirmed in the 1960s, when radio-active dating techniques were applied to rocks with different magnetism from around the world. They showed that the entire magnetic field of the Earth does reverse every few million years. Crucially, this is a truly global phenomenon; it applies everywhere on Earth at the same time.

Two researchers in Cambridge, Fred Vine and Drummond Matthews, combined the new magnetic evidence with Harry Hess's hypotheses to provide the clinching evidence for sea floor spreading. They said that molten rock wells out of the cracks in the sea floor at mid-ocean ridges, and spreads out evenly on either side. As it sets, the rock picks up a trace magnetism lined up with the Earth's magnetic field. After a time, the Earth's field reverses, so the magnetism being picked up by the setting rocks reverses. But this does not affect the magnetism of the rocks that have already set. The magnetic stripes should, if the idea is correct, form the same pattern on either side of a ridge, the corresponding stripes on either side of the ridge matching up with the record of Earth's magnetic reversals. In the mid 1960s, a survey carried out by a US research ship, the *Eltanin*, measured the pattern of magnetic reversals on either side of the ocean ridge in the Pacific, to the south of Easter Island. The measurements extended along a four-thousand-kilometre path, two thousand kilometres on either side of the ridge, and showed a perfectly matched mirror image. When the magnetic stripes from the *Eltanin* survey are plotted out on a chart, the paper can be folded along the line of the ridge, and the two patterns sit right on top of each other. There was no longer any room to doubt; sea floor spreading is real, and it explains continental drift.

There's a bonus. Because the pattern of magnetic reversals has been accurately dated by studies on land, the magnetic stripes on the ocean floor can be read like a bar code to reveal just when the rocks in each stripe were being laid down, from the present day at the ocean ridges to the oldest sea floor at the edges of the ocean basins. Once again, but more accurately than any previous estimates, the pattern shows that there is no sea floor more than 200 million years old.

The icing was put on the cake by Edward 'Teddy' Bullard, working at the University of Cambridge, in 1964. Bullard used a computer to calculate the best fit between the continents on either side of the Atlantic Ocean if the ocean were taken away. This jigsaw-puzzle-like fit was not made using the present-day coastlines, which depend upon sea level, but along the edges of the continental shelves, the true edges of the continents. In reality, the fit is not much better than you could make by cutting the shapes of the continents out from paper and shuffling them about on a table top. But electronic computers were still new enough in the mid 1960s for people (even scientists!) to be impressed by the apparent objectivity of the exercise, and as the other evidence in favour of sea floor spreading piled up, Bullard's map became an iconic image, a proof of the reality of continental drift.

All that remained was to explain *how* the process happened. The explanation developed with astonishing speed.

4

What goes up must come down

The discovery of sea floor spreading was only the beginning of an understanding of the behaviour of our dynamic planet. Since the Earth is not expanding – which has been proved by modern measurements, including data from systems, or GPS – the crust that is being created at ocean ridges must be balanced by crust being destroyed at the same rate somewhere else. That 'somewhere else' is in trenches where oceanic crust – that is, sea floor – slides under the crust of continents and back down into the interior of the Earth. In short, what goes up, *must* come down. Once the places where the oceanic crust does go down were identified, this quickly led to a complete theory of the changing Earth, known by the name 'plate tectonics'.

Downtrodden and uplifted

Curiously, some of the first evidence that oceanic crust is being pushed down in this way came from observations that the crust is being uplifted along the edges of some continents – because, as we now know, it is riding up over the back of the sinking sea floor. One of the first people to study this uplift scientifically was Charles Darwin, who gained fame as a geologist long before he turned his attention to the question of evolution. During his travels with Robert FitzRoy on the *Beagle*, Darwin spent more of his time on land than he did at sea, exploring the interior of South America while the *Beagle* carried out its tedious survey of the continent's coast. On these expeditions Darwin gathered a great deal of evidence that the land had been uplifted. Shells and shingle, as well as the fossilized remains of sea creatures, showed the presence of old beaches, and even former sea bed, far above present sea level. At one site in Peru hundreds of metres above sea level, there are even the remains of a whale. Rejecting the notion that this could all have been due to the effects of the biblical Flood, Darwin gradually began to wonder if even the mighty Andes could have been uplifted from the bed of a former ocean.

His ideas were crystallized early in 1835. He was ashore near the town of Valdivia, while the *Beagle* surveyed South America's western coast, when he experienced what seemed to be a major earthquake; it soon turned out to be only the fringe of the event. Shortly after Darwin came back on board the *Beagle*, the ship arrived at the city of Concepción, which had been destroyed by the earthquake and a tidal wave. Fresh mussel beds lined the shore, just above the high-water mark, a metre or so out of reach of the water. The mussels were all dead. Darwin realized that the land had been uplifted by this visible amount – more than a metre – during the course of a single earthquake which he knew, since he had experienced it himself, had lasted for no more than two minutes. He was convinced that over the span of geological time repeated earthquakes of this kind could indeed have raised the Andes to their present height.

The modern equivalent of Darwin's observations in South America comes from the western coastline of North America. In March 1964, the region near Anchorage, Alaska, was struck by an earthquake measuring 8.6 on the Richter scale (we discuss the Richter scale later, on page 85). This is big – probably comparable to the earthquake Darwin experienced in 1835. In Prince William Sound, in the fishing port of Cordova, the land was uplifted during the earthquake to such an extent that the harbour was left dry, and the fleet of boats used for salmon fishing was stranded. Investigating the phenomenon for the US Geological Survey, George Plafker found an exact analogy with the mussel beds studied by Darwin. Barnacles live only in the region between high and low tide, the intertidal zone, and cannot survive above the high tide mark. But Plafker found a band of dead barnacles almost two metres above the high tide line, indicating how much the land had been uplifted.

In the case of the 1964 earthquake, however, the opposite phenomenon had happened a little to the north. On an inlet roughly a hundred kilometres north-west of Cordova, the land subsided, and water flooded into the township of Portage, which had to be abandoned to the sea. But over the intervening decades, the region has silted up, leaving the ghostly remains of the wooden houses half-buried in the mud. When all the altitudes of the places affected around the region were plotted on a map, it showed that, along the edge of the continental shelf and the coast itself, the crust had been uplifted, while a little way inland and along the inlet it had subsided, as a piece of the continental crust had been tilted and distorted by pressure from the seaward side. This is how geologists learned that the floor of the Pacific Ocean is pushing up against and under the continental crust of Alaska. This process doesn't happen smoothly. Like trying to push a blunt carpenter's plane across a piece of rough wood, the movement is jerky. The two pieces of crust stick together while pressure builds up, and then, when it reaches a critical point, they jerk past each other and stick again. Earthquakes and uplift occur during the jerks, but at the same time the oceanic crust jerks forward and downward, under the continental crust. Exactly the same process explains what Darwin witnessed in South America.

How often does this happen? Plafker studied the island of Middleton, in the Prince William Sound, which has an unusual shape, akin to terracing. The island was uplifted during the 1964 earthquake, forming a new terrace a couple of metres above sea level. Plafker realized that the higher terraces had been formed in the same way, by repeated uplift associated with earlier earthquakes. By measuring the residual radioactivity produced by carbon in pieces of old driftwood from each of the terraces, he was able to work out that earthquakes like the 1964 event happen roughly every eight hundred years. If an uplift of two metres occurred during earthquakes spaced, on average, eight hundred years apart, then in a million years – the blink of an eye in geological time – the land would be uplifted by 2.5 kilometres, or 1.5 miles, nearly half the height of the Andes.

But by the second half of the 1960s it wasn't necessary to wait for hundreds of years to see the uplift occurring in major earthquakes. By then, instruments were sensitive enough to measure much smaller changes in the Earth's crust associated with much smaller earthquakes, and this fleshed out the global picture of how oceanic crust is swallowed up, beneath the continents.

The shaking Earth

Even before studies of uplift suggested a link between mountain-building and drifting continents, oceanographers had known about a series of deep trenches that run around most of the boundary of the Pacific Ocean. The trenches are like very deep valleys on the sea floor. The average depth of the sea is just 3.7 kilometres (2.3 miles), but a typical trench goes down six or seven kilometres – about four miles – below sea level; the deepest, a trench near the Mariana Islands, to the north of New Guinea, is eleven kilometres (seven miles) deep. These trenches always lie about a hundred kilometres, or sixty miles, out from the shore, and they usually run parallel to chains of volcanoes, which tend to follow curving lines called volcanic arcs.

In the 1950s, as seismologists began to study the pattern of small earthquakes around the Pacific, they discovered something else interesting about the trenches. Most of the seismic activity around the Pacific is, literally, around the edge of the ocean. What's more, it is concentrated in a layer of rock a few tens of kilometres thick that runs from beneath each ocean trench in a slope down into the interior of the Earth. These earthquake regions are called Benioff Zones, after the seismologist Hugo Benioff, who studied them. The slab of rock in which the earthquakes occur is part of the lithosphere, the rigid outer layer of the Earth which is made up of the crust itself and the part of the mantle, the next layer down, to which the crust is attached. Benioff said that the earthquake zones he had discovered must be places where the oceanic lithosphere is sliding down into the interior underneath a continent, producing a deep ocean trench in the process.

At first, this seemed an outrageous proposal. It was hard to believe the evidence that the crust of the sea floor is actually being destroyed at the edges of the Pacific Ocean. But the pattern of activity revealed by the 1964 Alaska earthquake changed geologists' minds. It turned out that wherever there is a Benioff Zone there are traces of large earthquakes and uplift like the events that occurred in Alaska that year, and which have been happening in that part of the world every eight hundred years or so.

So geologists had the answer to what happens to all the new sea floor that is being made from molten rock welling up from the mantle at spreading ridges. After travelling on the sea floor conveyor belt across to the edge of a continent, it is pushed downward and falls under its own weight deep inside the Earth, melting to become part of the mantle once again. Earthquakes occur where the lithosphere is distorted and bends on its way to the depths, and again at deeper levels where it runs into the mantle proper, creating a Benioff Zone. This whole process is called subduction, and the place where it happens is called a subduction zone.

Although subduction zones were first discovered along the edges of continents, there are also places where one piece of sea floor is sliding under another piece of sea floor. Where a subduction zone lies along the edge of a continent, the volcanic arc associated with the subduction zone forms a mountain chain that also lies along the edge of the continent – the volcanoes of the Andes are a typical example. But where one piece of sea floor dives under another piece of sea floor, the volcanoes appear as a series of islands along a curving path. These are known as island arcs; New Zealand is part of an island arc, which may be a modern-day example of how continents got started in the first place. Other island arcs include the Aleutians, the Kuriles, the Japanese islands, and the Philippines.

Volcanic refineries

But how, exactly, do volcanoes form? It turns out that one of the key ingredients is – water! The sea floor which is carried down into the depths of a subduction zone is no longer in the same simple state it was when it solidified at an ocean ridge. The rock itself has been modified by its interactions with water, which permeates through cracks in the rock and can get hot enough to boil, encouraging chemical reactions which change the composition of the rock. And at the edge of a continent the rock itself is covered in a layer of sediment, rich in the remains of organic life, washed down from the continent. As this mixture sinks, it gets hotter and is squeezed by extreme pressures – at a depth of fifty kilometres, or thirty miles, the pressure is fifteen thousand times the atmospheric pressure at the surface of the Earth, and the temperature reaches hundreds of degrees Celsius. One effect of all this is to squeeze water out of the rocks and into the mantle that surrounds the sinking slab of lithosphere. There, it has the same effect on the mantle that sprinkling salt on ice has on the ice – it encourages it to melt. Blobs of molten magma grow near the sinking slab of lithosphere, and because molten magma is lighter than solid rock, these blobs of molten material gradually rise upwards, slowly penetrating the crust and creating a chain of volcanoes above the sinking former sea floor.

While all this is going on, gases bubble out of the magma and eventually make their way to the surface, where they are released at volcanic vents. The most important of these gases are water vapour, carbon dioxide, and nitrogen. The nitrogen largely comes from the organic remains mixed in with the sediment carried downwards with the sinking slab of lithosphere. All of these gases are important to living things on the surface of the Earth, so life on Earth is intimately linked to the cycles involving the planet's non-living rocks.

There are other aspects of this volcanic refinery that are particularly important for human life. As the molten rock rises upward through the crust, it cools, and some of it solidifies on its way to the top. One of the standard ways in which geologists measure what is going on is in terms of the amount of silica that appears in different rocks. (Silica is just another name for the chemical silicon dioxide, and it is present in virtually all rocks on our planet.) The igneous rock basalt, which is released by volcanic activity, is by far the most important component of oceanic crust. It contains about fifty per cent silica (in terms of its weight). But the crust that forms the continents has a different composition, including sixty per cent silica. As a result, continental crust is lighter than oceanic crust, which is why sea floor dives under continents, but continents do not dive under sea floor.

Some rocks, of course, contain even more silica than the average – the metamorphic rock granite, for example, is about seventy-five per cent silica. The reason why continental crust contains relatively more silica than basalt is that other materials have been taken away from the mixture typical of basalt on its journey to the surface from a subduction zone. Those other things include minerals such as quartz, which crystallizes out of the molten magma, leaving it relatively richer in silica as it continues its rise, and compounds rich in metals such as iron, copper, silver, and gold, which solidify out higher up. This 'leaching' is the source of the wealth of precious metals found in South America and elsewhere, which have lured generations of explorers and bankrolled great empires.

Plate tectonics

All of these processes are now understood within the framework of plate tectonics, which explains sea floor spreading, subduction, mountain-building, and much more. 'Tectonics' means construction, so the name means 'constructing plates'; the same root gives us the word architect. The plates are the pieces of the Earth's crust that jostle about on the surface of our changing planet, taking continents with them for the ride.

The theory was mostly developed in 1967 and 1968, with leading contributions from the Canadian geologist Tuzo Wilson and the English geophysicist Dan McKenzie, and was given the name tectonics in a scientific paper published by McKenzie and his colleague Bob Parker in 1967. Although the idea has been refined and its details fleshed out since then, the overall theory has remained the same. The key is that instead of continents drifting through the rocks of the sea floor, they are carried around the globe on the back of giant plates, the jigsaw-like pieces of the Earth's crust. In the process, oceanic crust gets created and destroyed, but continental crust, once formed, stays, more or less indefinitely. As Harry Hess once put it:

> The ocean basins are impermanent features, and the continents are permanent, although they may be torn apart or welded together and their margins deformed. The continents are

carried passively on the mantle with convection and do not plow through oceanic crust.

This is plate tectonics in a nutshell.

The rigid outer shell of the Earth – the lithosphere – is broken up into seven large plates and several smaller plates. Some plates are made entirely of oceanic crust, but others are made up from both oceanic and continental crust. No plate can move independently, because whatever it does affects its neighbours, and whatever they do affects their neighbours, and so on. We've already come across two of the ways in which plates can interact with their neighbours. A spreading ridge is an example of a constructive margin, where new crust is being created. The activity associated with a deep ocean trench is an example of a destructive margin, where crust is being destroyed. On balance, crust is being created and destroyed at the same rate, when the whole planet is taken into consideration.

But plates can also rub past each other without crust being either created or destroyed. One place where this is happening today is in California, where the eastern edge of the Pacific plate is rubbing up against the North American plate, moving northward relative to the continent. This is responsible for all of the earthquake activity along the famous San Andreas Fault. In fact, the sliver of land to the west of the fault, including the finger known as Baja California, is riding on the Pacific plate and moving north; geologically speaking, it is not part of the North American continent. If the process continues, Baja California will end up becoming part of Alaska. This kind of plate boundary is called a conservative margin, because the amount of crust is conserved.

There's one last type of plate boundary, which is really a special case of a conservative margin. This is a collision margin, where two continents have come together as the sea floor between them has been swallowed up in subduction zones.

The result is a big crunch, a slow-motion collision in which the land is squeezed and pushed upward to make high mountain ranges, such as the Himalayas. Unlike the activity associated with subduction zones, there are no volcanoes among the mountains created in this way. So alongside its other successes, the theory of plate tectonics explains why there are volcanoes in the Andes but not in the Himalayas.

Interesting things also happen when a continent runs into a destructive margin – when the irresistible force meets the immovable object. This happens when the continent is riding on the back of a plate whose oceanic crust is being destroyed at a subduction zone, diving underneath another piece of oceanic crust. From the perspective of the continent, the destructive margin gets closer and closer as oceanic crust is destroyed; from the perspective of the subduction zone, the continent gets closer and closer. Either way, as the ocean floor shrinks away, there comes a time when the continent arrives at the subduction zone. But the subduction zone cannot swallow continental crust, and the global forces of tectonics may not allow the destruction of crust at the margin to stop. So a slab of lithosphere breaks away beneath the oceanic crust, and the oceanic crust from the other side of the margin is forced to dive down under the continent, with the destruction of crust continuing in the opposite direction.

Events like this can send shudders through the entire tectonic system, and can have repercussions on the other side of the ocean, or even on the other side of the world. One possible example of this kind of long-range connection can be seen in the Pacific Ocean, where the Hawaiian islands are part of a chain of more than 130 volcanic islands and guyots that runs all the way to the Aleutians of Alaska, over a distance of 5800 kilometres (3600 miles), with a distinct bend in the middle. South of the bend this is known as the Hawaiian chain, and north of the bend it is called the Emperor chain, but it is really one geological feature. This chain of islands likely formed via volcanism near the present-day location of the Big Island of Hawaii, where there is a hot spot in the crust, a plume of material from inside the Earth that is being carried upward by convection. This hot spot has persisted for tens of millions of years, and has punched a series of holes in the Earth's crust as the Pacific plate has drifted over its site at a rate of about ten centimetres per year. There are similar chains of volcanic islands, thought to have been made in the same way, in other parts of the world.

The youngest of the islands in the Hawaii-Emperor chain are found nearest to the hot spot, near Hawaii, while the oldest are the ones farther away, in the Emperor part of the chain. This neatly explains both parts of the chain in terms of the steady drift of the Pacific plate over a more or less stationary hot spot. (There is some evidence that the hot spot itself may have moved a little, but not enough to explain the whole chain.) But why is there a bend in the middle of the chain?

The best explanation is that tens of millions of years ago something gave the Pacific plate a jolt and changed the direction in which it was moving, directing it slightly more to the west. A clue comes from a careful investigation of the ages of rocks from the volcanoes around the bend. These turn out to be between fifty million and forty-two million years old, with the older rocks farther around the bend from Hawaii. In other words, whatever process it was that caused the change in direction began about fifty million years ago, and it took about eight million years for the plate to adjust and settle down into its new motion.

At exactly that time, fifty million years ago, there was an upsurge in volcanic activity far away on the western edge of the Pacific plate. This extended along 2200 kilometres (1400 miles) of the western edge of the Pacific plate, and is explained in terms of a developing subduction zone there. This allowed the Pacific plate to adjust its motion in a more westerly direction, exactly what is required to explain the bend in the Hawaii-Emperor chain. This may have marked the beginning of the end for the

Pacific plate, for if the subduction along the western margin continues, eventually the whole Pacific will disappear, and North America will collide with Asia. The best example of what happens in such a collision can be seen today in a region we have already mentioned – the Himalayas, one result of a collision between India and the landmass to the north. It is even possible that the collision between India and Eurasia – two landmasses that became locked in a firm embrace about fifty million years ago – was itself part of the global shake-up that triggered the subduction along the western margin of the Pacific.

Death of an ocean, birth of a mountain range

Some 200 million years ago, all of the landmasses of the Earth were linked together in the single supercontinent called Pangaea. (As mentioned in chapter 3, meteorologist Alfred Wegener chose the name because it meant 'whole Earth'.) When Pangaea began to break up, between 150 million and 100 million years ago, the landmasses that would become North America and Eurasia split off first, leaving behind the land that would become South America, Africa, Antarctica, Australia, and India still welded together as a single southern continent, Gondwana. To the west, there was never much of a gap between Gondwana and North America, but to the east a great ocean, called the Tethys Ocean, opened up between eastern Eurasia and Gondwana. As Gondwana began to split up, Africa and South America separated from the continent, and India started its independent life as a block of continental material just to the east of the southern part of Africa. The last piece to separate was Australia, leaving Antarctica as Gondwana's final remnant.

As Africa moved closer to Eurasia in a northward drift, it soon closed off the western end of the Tethys Ocean. But for tens of millions of years the eastern part of the ocean remained in the narrowing gap between India and Eurasia. Although India started out at the same latitude as southern Africa, it moved north a lot faster than Africa did. About eighty-five million years ago, India was still far to the south of Eurasia, with the shrinking Tethys Ocean in between. But between about eighty-five million years ago and fifty million years ago India moved northward at a rate of about ten centimetres, or four inches, per year, a very high speed for continental drift, covering a distance of about 3500 kilometres, or 2200 miles, which is roughly the same as the width of Australia.

When India slammed into Eurasia, the Tethys Ocean had shrunk almost entirely out of existence, leaving just a small portion to the west, north of Africa and south of Europe – the Mediterranean Sea. In another classic example of the irresistible force meeting the immovable object, something had to give, and that something was the immovable object – Eurasia itself. As the ripples from the impact spread around the planet and played a part in the development of subduction along the western Pacific (and creating the bend in the Hawaii-Emperor chain), the movement of India north was slowed down to about five centimetres per year, but did not halt. (That slowing rate of movement is still twice as fast as the rate at which the Atlantic Ocean is widening today.) Sediments that had once lain at the bottom of the Tethys Ocean were squeezed and pushed upward by the relentless pressure from India, forming the high Himalayas, mountains that are still growing today. Every day, Mount Everest is a little higher than it was the day before; so every mountaineer who reaches the top of Mount Everest and stands there, on the former floor of the Tethys Ocean, has the right to claim, until the next person comes along, that he or she is the only person to have reached the highest point on Earth. The relentless pressure from India has also extended beyond the Himalayas, to create the high plateau of Tibet.

This process can only be explained by the discovery of a curious property of the rock from which continents are made.

Although it seems solid enough to us, if this rock is pushed firmly but steadily (pushed very hard!), it flows like a liquid. Strange though this sounds, there are things that behave a bit like this in everyday life. Some household paints form a jelly-like solid in the can, but when you dip a brush in and spread the paint on the wall it flows like a liquid. And even ordinary custard, the kind you put on a sponge cake, has similar properties. It flows like a liquid most of the time, but if you make a big vat of custard it is possible to run across it without sinking, as long as you keep moving (we've seen this done, although we haven't actually done it ourselves).

The thing about a fluid is that it will flow, and continental crust will flow, under the right conditions. The right conditions are a very firm, steady pressure – which is just what India has been applying to Asia for fifty million years. As India has pushed onward into Asia, the region beyond the Himalayas grew to the maximum height that could be supported by Asia's crust. Beyond the Himalayas, the crust was raised as high as it could go without flowing, and then flowed out in front of the advancing mountains to make the high plateau of Tibet. But it didn't happen quietly.

We said that there are no volcanoes in the Himalayas, and that's true. But there are signs of recent volcanic activity on the Tibetan plateau, and there are steep valleys, running north–south across the plateau, that mark the sites of recent cracks in the Earth's crust. How can these have formed in a region of fluid flow, which ought to smooth out irregularities? The answer lies deep beneath the surface, in the 'roots' of the mountains themselves.

When a piece of the Earth's crust is squeezed by tectonic forces, it gets pushed both upwards and downwards as it is squashed. Mountain ranges have roots extending far below the surface, just as icebergs have roots extending deep below the water level. The root is formed from the rock of the lithosphere, the cold top layer of the mantle. When the high plateau of Tibet formed, it also had a root of this kind, reaching down into the asthenosphere, the next layer of the mantle. The asthenosphere is much hotter than the lithosphere, so the cold root of the Tibetan plateau warmed up and essentially melted away, dropping off the bottom of the lithosphere. Released from the weight of this root, the Tibetan plateau bobbed upwards by a couple of kilometres, while molten material from the asthenosphere flowed into the gap left by the root; and this molten material eventually came to the surface via volcanoes. But the plateau bobbed up too far to be supported by the fluid crust and sank back downwards again, spreading out and cracking to create the valleys seen in Tibet today.

Thus the presence of hot springs on the high plateau of Tibet can be directly linked to the bend in the Hawaii-Emperor chain, far away in the middle of the Pacific. There are few better examples of the power of plate tectonics theory to explain how the world has got to be the way it is today. Plate tectonics can also explain the most important feature of the world, as far as land-based animals like us are concerned – the fact that there is any land at all.

Making and breaking continents

Although sea floor crust is thinner than continental crust, it is also stronger. Sea floor is made up of solid slabs of volcanic rock that have set as they poured out of ocean ridges; but continental crust is made up of all kinds of bits and pieces that have been broken up and stuck back together. That's why continental crust can be distorted, and flow to make features like the Tibetan plateau. Geophysicists sometimes compare this structure to the contents of a bag of flour. The flour is made up of a large number of solid grains. It is certainly not a liquid, but because the grains are small the flour can be poured out of the bag like a liquid. In the same way, it's relatively easy to distort the continental crust because it is made up of fragments, which can slide past one another – though not as easily as the much smoother grains of flour do so. This fragmented structure was present even in the very first continental crust that formed.

Just after the Earth cooled, there was no continental crust at all. The first solid surface of our planet was like a fiery version of the pack ice we mentioned earlier, with slabs of solid 'magmatic' rock floating on a sea of molten rock, which bubbled from the cracks between these early plates. Something similar can be seen today, on a much smaller scale, in the crater of a volcano; the classic example, beloved by geophysicists, is Erta Ale, in Ethiopia. When the Earth cooled further and water condensed out of the atmosphere to fall as rain, it covered the entire world with a uniform ocean. Land began to poke itself above the surface of the world's ocean as some of the jostling plates, carried around by convection currents in the fluid magma beneath, began to push under the adjacent plates, creating the first subduction zones and the first chains of volcanic island arcs, made from igneous, volcanic rock like andesite.

But from the very beginning, this wasn't the only way that continental crust accumulated. When one plate pushes under another, even if both of them are made of oceanic crust, bits of rock get shaved off, just as a carpenter's plane scrapes off shavings from a block of wood. These rock shavings also stuck to the growing landmass, and as proper continents began to grow the mixture of material accumulating around their edges was enhanced by sediments washed down from the land. Eventually, tectonic activity brought pieces of crust formed in this way together, and welded them into larger continents, just as we see happening with India and Asia today.

It's impossible to tell exactly how long it took to produce landmasses the size of the present continents, because we don't know how fast the tectonic processes went on when the Earth was young. But a reasonable guess is that about half of the continental crust that exists today, and perhaps a bit more, had already been formed by about 2.5 billion years ago. Geological evidence for the rate at which the Andes, whose activity so impressed Charles Darwin, have grown over the past fifty million years suggests that overall continental crust is still being created today, at a rate of about one cubic kilometre per year. Not much in a single year, but this makes sixty-five million cubic kilometres, or fifteen million cubic miles, in the time since the death of the dinosaurs.

Overall, this is a very satisfactory explanation of where the continental crust came from. Yet, it raises a question. If these processes have been going on for billions of years, why hasn't all the continental crust been welded together into one supercontinent? Putting it another way, since it *was* all welded together in Pangaea not so long ago, why did the supercontinent break up? The answer is probably linked, once again, to the nature of the Hawaii-Emperor chain and the hot spot beneath Hawaii.

We now know that a hot spot like the one that formed the Hawaii-Emperor chain is just the tip of a bigger, deeper site of activity, called a mantle plume. All around the Hawaiian island region, the sea floor is about 1.5 kilometres – a mile – less deep than it is across most of the Pacific Ocean. This is because the sea floor has been pushed up into a dome, roughly a thousand kilometres across, by this plume, welling up beneath it. The hot spot itself is like a pimple at the centre of this dome, releasing a quite modest amount of volcanic material at the surface. The sea floor around Hawaii is strong enough to keep a lid on the plume itself. But there are two kinds of places where weaker crust could allow much more of the mantle plume's material to break through to the surface.

The first is at an ocean ridge. Indeed, Iceland is a piece of sea floor, complete with an ocean ridge running through it, that has been raised up in this way by the presence of a plume beneath it. Iceland belongs at the bottom of the ocean, but it is 2.5 kilometres (1.5 miles) higher than the rest of the Mid-Atlantic Ridge. It may seem a bit of a coincidence that a plume should lie exactly under a spreading ridge in this way, but the most likely explanation is that a weakness in the crust allowed the plume to rise to the surface, and the resulting activity extended the crack in the crust, creating the ridge and spreading system that made the Atlantic Ocean. One intriguing but unproven suggestion is that Iceland marks the spot where a large meteorite struck the surface of the Earth, punching a hole in the crust, which allowed the magma to well up and triggered the whole process.

The other place where an upwelling plume can break through to the surface with relative ease is under the crust of a continent, which is weaker than the crust of the ocean floor. This would split the continent apart, and start the processes that make the continents drift away from each other. When the plume broke through to the surface entirely, it would release huge amounts of molten rock, spreading over the surface and solidifying into huge quantities of flood basalts. Just such a layer of basalt, more than a kilometre thick, covers much of western India, and is known as the Deccan Traps. At its thickest, the layer of basalt is three kilometres (1.8 miles) in depth, and it covers an area of 550,000 square kilometres (200,000 square miles). It formed within a span of about a million years, some sixty-five million years ago. Unluckily for India, on its way north it seems to have drifted over a mantle plume, which split the continent in two and left a large part of the original landmass as an underwater bank of basalt to the south-west of India's present location, between the Seychelles and the Comoros islands.

Similar basalts are found both on the eastern side of South America (in Argentina, Paraguay, and Brazil) and on the western side of Africa (around Angola). These have all been dated, on both sides of the Atlantic, to around 120 million years old; they are evidence of the powerful forces associated with a mantle plume that ripped the two pieces of crust apart. Older floods of this kind of basalt, found in places such as Siberia, where there are traps 250 million years old covering 300,000 square kilometres (100,000 square miles) of the Arctic, tell us that there have been previous episodes of continental break-up in the long life of the Earth, before the continents came together to form Pangaea. And the process is still going on today.

Breaking up is not so hard to do

The after-effects of mantle plume activity can be seen today in the region of the Red Sea, the Horn of Africa, and the African Great Rift Valley. There, about forty million years ago, a mantle plume rose nearly to the surface, creating a dome in which the rock cracked and volcanic lava flooded out over the crust. As the pressure was released, the plume collapsed back into the interior of the Earth, and the crust around the centre of the dome collapsed downwards, creating a series of deep rifts in the Earth's surface. One of these rifts in the side of the dome became the Great Rift Valley; nearby, another rift dropped below sea level, forming the Red Sea as water rushed in to the resulting crack. At the same time, molten material continued to well up into the crack, which is about 1800 kilometres (1100 miles) long, pushing the two sides farther apart. The Red Sea is an ocean in miniature, complete with a mid-ocean ridge; it has been widening for about ten million years, at a rate of about eighty millimetres (three inches) per year, splitting Africa away from Asia.

At the southern end of the Red Sea, the rift meets the northern end of the Great Rift Valley, and the western extension of a sea floor ridge running through the Gulf of Aden north of the Horn of Africa and linking the system to the spreading ridges of the Indian Ocean. The three rifts together form a kind of 'Y' shape, which is typical of the pattern of rifting that occurs when the dome above a mantle plume cracks. But, as in this case, not all of the three rifts necessarily turn into spreading ridges immediately.

The Great Rift Valley is one of the longest fault systems on land in the world, and may yet split Africa in two – the Afar

Depression in Ethiopia is already well below sea level, and in 2006 it was rocked by a series of earthquakes in which the depression widened by about three metres (ten feet) and sank by a further one hundred metres. Before too long, perhaps within a hundred million years, if this kind of activity continues the sea will flood in, eventually filling the entire Great Rift Valley. It will be an historic loss, not just because of the size of the system and its geological significance, but because the Great Rift Valley has a special place in the story of the relationship between people and our home planet.

From the Red Sea, the rift extends over a distance of some five thousand kilometres, or three thousand miles, south to Mozambique and the mouth of the Limpopo River. Although the average width of the valley is just fifty kilometres (thirty miles), in places it is several hundred kilometres across, with steep-sided walls typically just under a kilometre high, but in places soaring above 2.5 kilometres, or 1.5 miles high. The edge of the rift is dotted with volcanoes, including Erta Ale in Ethiopia, Mount Kenya, and Kilimanjaro; the floor of the valley contains very many deep lakes, such as Lake Nyasa and Lake Turkana. This unusual geography has created conditions in the valley where life has flourished, while the ash from volcanic eruptions has helped to preserve the remains of living things in the valley for millions of years, as fossils. Because the fossils have been preserved, we know that the special living conditions made the region of the rift around Kenya and Ethiopia the cradle of humankind, where many ape-like creatures flourished and one branch of the ape family evolved to become our immediate ancestors (see chapter 10). So there is a direct link between the forces that tear continents apart and the existence of *Homo sapiens*. Which means there is a direct link between our existence and the forces that operate deep inside the Earth.

5

Inside and outside the Earth

It's sometimes said that we know more about the stars than we do about the interior of our home planet. That's probably true – after all, we can actually see the stars, and analyse their light using techniques such as spectroscopy, which measures the energy or frequency of the starlight. But we cannot see what lies deep beneath our feet. So almost everything scientists know about the inner structure of the Earth, and what goes on there, has been gained by studying the way vibrations caused by earthquakes travel through our planet.

Many popular accounts make this science sound quite straightforward, saying that studying earthquake vibrations reveals the inner structure of the Earth much in the way that an X-ray or a CAT scan can reveal the inner structure of your body. But Teddy Bullard, the eminent Cambridge geophysicist who produced the first computer fit of the continents on either side of the Atlantic, told us once over coffee in Cambridge that trying to work out the inner structure of the Earth by studying the vibrations from earthquakes is more like trying to work out the inner structure of a grand piano by listening to the noise it makes when it is pushed down a flight of stairs. So it's really much harder than analysing an X-ray.

Scanning inside the Earth

The initials CAT stand for 'computer axial tomography', and the term tomography is derived from the Greek word *tomos*, meaning a section, as in 'cross-section'. Earth scientists have borrowed the jargon and rather ambitiously given the name seismic tomography to their studies of the Earth's interior, even though the pictures they get will never be as good as CAT scans.

Shock waves from earthquakes – seismic waves – travel through the interior of the Earth at different speeds depending on what kind of rock they are travelling through. Among other things, the speed at which they move depends on the temperature of the rock, and whether that rock is soft or hard. As the waves travel through different kinds of rock, or the same kind of rock at different temperatures, the direction they are moving in can change, a bit like the way the direction of a beam of light can change when it moves from one kind of material (such as air) into another kind of material (such as glass, or water). This is called refraction. Also, when a seismic wave travelling through one kind of rock arrives at a boundary with a different kind of rock, it can be reflected, like light being reflected from a mirror.

This would already be enough to provide valuable information about the interior of the Earth, but as a bonus there are two different kinds of waves to study. One kind, called pressure waves (or just P-waves), are like sound waves, and move with a push-pull motion. They are like the waves you can make with the toy called a Slinky, which is like a long spring. The other kind are called shear waves (or just S-waves), because they move with a to-and-fro sideways motion, like a snake, or the waves you can make by sending ripples running along a rope. Since P-waves travel faster than S-waves, they arrive first at detectors, called seismographs, or seismometers, and for that reason the initial P in P-wave is sometimes taken to mean 'primary'. Shear waves arrive second, so the S in S-wave can also stand for 'secondary'. In the body of the Earth, P-waves travel at speeds between about seven kilometres per second and fourteen kilometres per second (four miles per second and eight miles per second), while S-waves travel at speeds between about four kilometres per second and eight kilometres per second (2.5 miles per second and five miles per second). As a rule of thumb, in a given kind of rock, the S-wave travels at sixty per cent of the speed of the P-wave.

P-waves can travel through both liquids and solids, but S-waves cannot travel through liquids. It was the discovery that P-waves can travel through some regions of the interior of our planet where S-waves cannot travel that revealed the molten outer core of the Earth.

There are also surface waves, which, as their name implies, travel across the surface of the Earth. These can be very powerful and do a great deal of damage, but they do not tell us much about the deep interior of the Earth. Because they do probe the deep interior, together P- and S-waves are known as body waves.

Of course, if you only had one seismometer with which to study earthquake vibrations, none of this would tell you much about the deep interior of the Earth. You wouldn't be able to make much more sense of them than a person would be able to make sense of the noise made by a grand piano being pushed down a flight of stairs. But there are hundreds of sensitive seismometers, linked together in networks that are spread over a large part of the surface of the Earth, and every day there are many small earthquakes going off somewhere, creating vibrations that can be picked up by those instruments and analysed. Once again, geophysicists owe their understanding of the Earth in no small measure to the Cold War. After the testing of nuclear bombs in the atmosphere was banned in the early 1960s, the military establishment and governments of the superpowers wanted to monitor the underground tests being carried out by their rivals. This led to the establishment by the US government of the Worldwide Standardized Seismograph Network, or WSSN, which gathers data from seismic stations all over the world to be analysed at a central laboratory in the United States. This is still the most important network of this kind.

Global networks are good at giving a broad picture of the structure of the Earth, especially its division into different layers. But there are also networks where many seismographs are placed more closely together to get a more detailed picture of what is going on in one particular region of the globe. The smallest details they can 'see' are on a scale of a few kilometres. Together, these techniques give geophysicists a picture of what is happening deep beneath our feet.

Layers within layers

Beneath the crust, which is a bit like the shell of an egg, the interior of the Earth resembles the interior of the egg itself, with a core (corresponding to the yolk) surrounded by a deep layer of material (corresponding to the white of the egg). But unlike the yolk and white of an egg, these main layers of the Earth's interior are each divided into an inner region and an outer region, identified from seismic studies.

Starting at the surface and working downwards, towards the centre of the Earth, the crust is, on average, about seven kilometres (four miles) thick under the oceans and about thirty-five kilometres (twenty miles) thick on the continents. It makes up just 0.6% of the volume of the Earth, and 0.4% of its mass. The base of the crust is marked by a boundary called the Mohorovičić discontinuity (or Moho), after the Croatian scientist who discovered it. The main layer below the crust is called the mantle, and is divided into two parts, the upper mantle and the lower mantle.

But the transition from the crust to the mantle is not a neat dividing line. The top of the mantle is the solid, rocky region that, together with the crust, is called the lithosphere; it extends down to a depth of about 250 kilometres (150 miles) below the continents, but is much thinner under the oceans and little thicker than the crust itself at mid-ocean ridges. Just below the lithosphere, there is a semi-liquid region, a little more than 100 kilometres (sixty miles) thick, still chemically part of the mantle; this is the asthenosphere. This part of the mantle is the key to plate tectonics. Because the asthenosphere is semi-liquid, the solid lithosphere above, including the crust, can slide about on it, allowing plates to move, the sea floor to spread, and continents to drift. The plates that are the feature of plate tectonics don't just consist of crust, but are slabs of rock that include both crust and the top of the upper mantle. Including the lithosphere and the asthenosphere, the upper mantle extends down to about 670 kilometres (415 miles) from the surface, and the lower mantle goes down to about 2900 kilometres (1800 miles). It makes up eighty-two per cent of the volume of the Earth, and sixty-seven per cent of its mass. P-waves travel at about eight kilometres per second in the top of the mantle, and at nearly fourteen kilometres (8.5 miles) per second at its base.

At this depth, there is a much more dramatic transition to a region of liquid material, through which S-waves cannot pass, called the outer core. At the base of the outer core, about 5100 kilometres, or 3100 miles, below the surface of the Earth, we reach the top of the inner core, a solid lump of material some 2400 kilometres, or 1500 miles, across, about two-thirds the diameter of our moon. One curious feature of the solid inner core is that it is rotating slightly faster than the rest of the Earth, and has gained one-tenth of a rotation in the past thirty years, a rate of a little more than one degree per year – that is, 1/360 of a circle. The whole core is almost exactly the same size as the planet Mars. But both parts of the core, the seismic studies reveal, are much more dense than the mantle above. Altogether, the distance from the surface to the centre of the Earth is 6371 kilometres, or 3959 miles.

All of these distances are averages. The depths of the various boundaries are slightly different in different places, and may change as time passes. In particular, the solid inner core is thought to be growing slowly, as part of the liquid outer core crystallizes on top of it. This is one source of the energy that keeps the interior of our planet hot (along with radioactive energy), because when liquids solidify they give out energy, called latent heat. So far, about four per cent of the core has crystallized, and it will take another four billion years or so for the rest to solidify. The whole core occupies only 17.4% of the volume of the Earth, but contains 32.6% of its mass, an indication of its very high density compared with the rest of our planet – about twelve grams per cubic centimetre (0.4 pounds per cubic inch), which is twelve times the density of water, and a little more than the density of lead in any of the forms in which you may find it at the surface of the Earth.

Mixing the mantle

Plate tectonics is driven by convection in the mantle – or mantles. This sounds crazy: didn't we just say that the mantle is solid rock, through which S-waves can travel? But consider the rocks of the Tibetan plateau, which seem solid to us but can flow very slowly. This is another example of the same sort of thing. If the mantle is shaken suddenly by an earthquake, vibrations can travel through it, like sound waves travelling through a ringing bell. But push it steadily for a long time and it will gradually flow like a very sticky liquid.

There is actually something we use in everyday life that behaves like this – glass. If you hit a piece of glass with a hammer, it will shatter, like a solid. But if you look carefully at the stained glass in old cathedral windows, you will see that it is thicker at the bottom than at the top. This is because, over the centuries since the glass was installed, it has flowed slowly downwards, like very sticky syrup, under the pull of gravity. The mantle is like glass, but even stickier.

Because of the division of the mantle into upper and lower regions, separated by a narrow boundary that has been revealed by seismic surveys, nobody is quite sure how this convection works. One possibility is that the two parts of the mantle each have their own convective system, like a double boiler, one on top of the other. The opposite extreme would be if the whole mantle convected as a single unit, more or less ignoring the boundary between the upper and lower mantle.

But the best explanation seems to involve a bit of both of these ideas. Slabs of sea floor that have been forced downward by subduction can be traced by seismic scans all the way down to the boundary between the upper and lower mantles at a depth of about 650 kilometres (400 miles). There, they seem to spread out in a layer that collects for perhaps hundreds of millions of years, before it piles up so much that it breaks through the boundary and carries on into the lower mantle, like an avalanche, almost all of the way down to the top of the outer core. There is some seismic evidence that solid slabs of material accumulate at the base of the mantle, where it may take as long as a billion years for the heat from the core to warm them sufficiently to make the material start rising back towards the surface. The re-heated material then becomes part of a rising plume that breaks right through the boundary into the upper mantle and forms a hot spot – like the one under Hawaii, and the one that is cracking the African continent apart. The heat that drives these plumes upward and is the primary cause of sea floor spreading comes all the way from the bottom of the mantle, lifting a column nearly three thousand kilometres, or 1800 miles, high.

All this is possible only because the interior of the Earth is hotter than its surface. But the chemical composition of the hot rocks, and a rather surprising lubricating agent, are also needed to explain the way plates move.

Hot rocks and chemical cookies

If you drill into the Earth's crust, or dig a deep mine shaft, you find that the deeper you go, the hotter the Earth gets. The deep gold mines of South Africa reach five kilometres (three miles) down into the crust, and the temperature of the rocks they penetrate increases by between ten and fifteen degrees Celsius for every kilometre below the surface (the equivalent of eighteen and twenty-seven degrees Fahrenheit). This is called the temperature gradient. If this gradient continued all the way to the centre of the Earth, the temperature at the centre would be about 80,000 degrees Celsius, or 144,000 degrees Fahrenheit. But this calculation would only be appropriate if there were no convection occurring in the mantle. Convection changes the way heat is transported from the interior to the surface, and reduces the increase in temperature with depth.

Convection is a circulating system in which fluid is heated from below and the hot fluid rises in columns, surrounded by cold fluid sinking back down into the depths. This happens because the hot fluid at the bottom expands and becomes less dense, so that it rises, while the cold fluid at the top, which has radiated or conducted its heat away upwards, contracts and becomes more dense, so it shrinks. Convection is a very efficient way of transporting heat from the depths up to the surface, and this process tends to smooth out any temperature differences between the top and the bottom of a fluid. The convection in the mantle has been likened by geophysicists to convection in a cauldron of hot tar, and it certainly reduces the temperature difference between the surface and the core, compared with the estimate that can be extrapolated from mine shaft measurements. But in order to find out how much the temperature gradient is reduced, we need to know what the core is made of.

We know a great deal about the composition of the oceanic and continental crust, because we can easily get hold of pieces of their rock for analysis. Continental crust is rich in minerals such as quartz and feldspar, while oceanic crust, beneath its thin coating of sediment, is made of igneous rock, which cooled as it flowed in a liquid state from cracks in the Earth's crust. We even know a surprising amount about the composition of the mantle. For instance, diamonds are a form of carbon that can only be produced under conditions of extreme pressure, and diamonds are recovered today in material that erupted from volcanoes long ago. The presence of diamonds indicates that the rocks in which they are found were produced from volcanic material that originated at a depth of several hundred kilometres below the Earth's surface, in other words, in the mantle. These rocks often contain minerals such as olivine, garnet, and pyroxene, which together form a granular, igneous rock called peridotite – what the mantle is made of. That is the limit of our direct investigation of the composition of the interior of our planet. To learn more, we need a combination of chemistry, astronomy, and calculation.

Chemistry is involved in understanding both the top and the bottom of the mantle. In the previous chapter, we mentioned the importance of water as a lubricant in subduction zones, and its role in encouraging volcanoes to erupt. It may also be the crucial ingredient that makes the asthenosphere slippery, and that allows the Earth's plates to move at all. Water weakens rocks and encourages them to melt, like antifreeze poured on ice, and it also oozes into the cracks where two plates slide past one another, acting precisely as a lubricant, like oil. But that isn't the whole story. Water in the cracks between the plates also reacts with the material of the rocks themselves, making new minerals that crystallize out along the plates' edges. These minerals are weaker than the rest of the plates, which again encourages the plates to slide past each other. Without water, probably we would not have plate tectonic activity on Earth at all, and the lack of water is probably one reason why there is no plate tectonic activity on Venus, which is in many ways so nearly the Earth's twin.

In the core, water does not play a key role, except by its absence. Absence, though, turns out to be the key to our unlocking the core's composition.

Astronomers have identified meteorites – chunks of rock from space that have fallen to Earth – as leftover pieces of the ring of material that circled the sun and from which the planets formed. These pieces of rock are particularly rich in four elements – iron, oxygen, magnesium, and silicon – as well as an alloy of iron and nickel. If the Earth were made of the same mixture of material, in the same proportions, it would have exactly the measured density of our planet, which is 5520 kilograms per cubic metre, or 0.2 pounds per cubic inch. But if you took away the iron and nickel from a meteoritic mixture the size of the Earth, you would be left with a volume that is the same as the Earth's mantle, and a chemical composition that is almost identical to the olivine and pyroxene actually found there. And the amount of iron 'missing' from the mixture is exactly the amount needed to make the core of the Earth. This very simple calculation tells us that the core of the Earth must be mostly made of iron, with some nickel too.

Knowing the elements of the core allows you to work out its temperature, because we have evidence that the inner core is solid while the outer core is liquid. In such a case, the temperature at the boundary between the inner core and the outer core must be at the melting point of an iron-nickel mixture at the pressure corresponding to a depth of 5100 kilometres (3100 miles) below the surface of the Earth, where iron is squeezed to a density twelve times the density of water. Laboratory experiments have revealed that this temperature is roughly five thousand degrees Celsius, or ten thousand degrees Fahrenheit, which, as it happens, and entirely by coincidence, is very nearly the same as the temperature at the surface of the sun. Because a solid lump of iron and nickel is a very good conductor of heat, we can be sure that this is pretty much the temperature across the entire inner core. So we can say that the temperature at the centre of the Earth is about five thousand degrees Celsius.

The structure of the iron-nickel core, with a solid inner core surrounded by a liquid outer core, is responsible for one of the most important features of the planet from the point of view of life on Earth – its magnetic field. Nobody knows exactly how the magnetic field is generated, but it must be a result of physical currents of material, circulating in the electrically conducting outer core and generating magnetism as they do so, like a dynamo. Both calculations and experiments show that this kind of thing will not work with a completely liquid core, where fluid would circulate more evenly. The best explanation for the Earth's magnetic field is that the solid core is surrounded by cylinders of swirling material in the outer core, like a tennis ball surrounded by a ring of fat marker pens standing on their ends. A combination of convection and the twisting forces caused by the Earth's rotation (the Coriolis effect) generates a magnetic field in each 'marker pen' that contributes to the overall magnetic field. But in four billion years' time, when the entire core has solidified, the Earth will lose its powerful magnetic field.

The magnetic field actually marks the outer boundary of the Earth, in terms of its impact on the local space environment. So the centre of the Earth is directly connected to its outermost shell. Between the solid surface of the Earth and the outer reaches of its magnetic influence, however, there is a region that is essential for our own existence – the atmosphere.

Earth's blanket

For convenience, scientists divide up the atmosphere of the Earth into notional layers. The layers can be most easily described in terms of the way the temperature of the atmosphere changes with altitude, but the boundaries between these layers are not as distinct as might be suggested by the labels assigned by atmospheric scientists. In reality, the boundaries are always more or less indistinct, with gases mixing both upwards and downwards across the boundaries.

Air close to the surface of the Earth is warm, because the surface itself (both land and sea) is warmed by sunlight. Most of the energy in sunlight passes through the atmosphere without being absorbed, so it does not heat the atmosphere directly. The energy of sunlight warms the surface of the Earth, and this warms the air above it both directly, by the process of conduction, and, much more importantly, by radiating heat back out towards space. Convection also carries warm air upwards from the surface of the Earth, obeying the dictum 'hot air rises', before it cools and falls back down. This stirring of the atmosphere by convection is the reason we have weather. But the reason why air at the surface of the Earth is so warm is primarily because the radiation from the planet's surface exists at much longer wavelengths than visible sunlight, in what is called the infrared part of the spectrum. (The difference in wavelengths is a result of the simple fact that the surface of the Earth is much cooler than the surface of the sun.) Infrared radiation *is* absorbed by gases in the atmosphere, in particular by carbon dioxide and water vapour.

This process of warming the air near the surface of the Earth is often called the greenhouse effect, although in fact it is not the way the air in a greenhouse is kept warm. A real greenhouse traps heat inside because the glass roof acts like a lid and stops convection (hot air rising) so all the hot air stays in the greenhouse.

But there is no lid on the atmosphere close enough to the ground to explain why the surface of the Earth is so warm.

The strength of this natural greenhouse effect can be seen by comparing the temperature at the surface of the Earth with the temperature at the surface of the moon. The moon has no atmosphere and is, essentially, at the same distance from the sun as the Earth is. On the moon, averaging over the whole surface (both day and night sides), the temperature is about minus eighteen degrees Celsius, or zero degrees Fahrenheit; on Earth, the average planetary temperature is about fifteen degrees Celsius, or fifty-nine Fahrenheit. The natural atmospheric greenhouse effect keeps the surface of the Earth some thirty-three degrees Celsius warmer than it would be if there were no atmosphere.

Because the warming of the atmosphere comes from the surface of the Earth, the air is cooler farther from the surface. Air temperature falls up to an altitude of about eleven kilometres (6.8 miles), where the temperature is roughly -60 degrees Celsius (-75 Fahrenheit). The 'about' is a particularly vague expression in this case, since cooling stops at an altitude of roughly eight kilometres (five miles) above the poles, but as high as sixteen kilometres (ten miles) over the equator. The layer of air below this boundary is called the troposphere, and the boundary itself is called the tropopause, which marks the beginning of a region in which air temperature holds steady, then increases with altitude, following the maxim 'heat rises'. The blanket of air gets thinner with altitude, because the gravity of the Earth holds most of the bulk of the atmosphere close to the surface. For that reason, more than eighty per cent of the mass of the atmosphere lies in the troposphere.

Because of a combination of the thinness of the air and the cold, the tropopause defines the extreme upper limit of the habitable zone of the Earth. The air at the summit of Mount Everest, with a height of just under nine kilometres (5.5 miles), is just about breathable. If the lower limit of the life zone is set as the sea floor at a maximum depth of about eleven kilometres (seven miles), that means that the life zone of the Earth is just about twenty kilometres in thickness, compared with the 6371-kilometre (3958-mile) radius of the Earth. Remember, this is less than the comparative thickness of the skin on an apple compared with the size of the apple itself.

Because the air above the troposphere is warmer than the air below, the tropopause acts like a lid on the troposphere, preventing convection from rising farther into the atmosphere. But as we explained above, the lid here is much too far above the surface, and the air much too cold, to explain the warmth near the Earth's surface in terms of the process that keeps a genuine greenhouse warm inside. The troposphere is also the weather layer of our planet, the place where the weather machine of convection does its work.

Earth's sunshield

The warming layer of atmosphere above the tropopause is known as the stratosphere, and although there is no life there, the stratosphere is essential for the well-being of life forms on the surface of the Earth. Up to about fifty kilometres (thirty miles) above the surface, temperature increases with altitude, until at that height the air, although very thin, is nearly as warm as the air at sea level. The warming zone (which defines the stratosphere proper) extends from about fifteen kilometres, or nine miles, to about fifty kilometres in altitude, where there is a boundary, called the stratopause, above which the atmosphere cools once again.

The stratosphere is warm because, unlike the troposphere, it absorbs energy from incoming solar radiation. That is why, unlike the troposphere, it is hotter at the top (about ten degrees Celsius, or fifty degrees Fahrenheit) than it is at the bottom. The radiation that is absorbed in the stratosphere is in the ultraviolet part of the spectrum, with wavelengths shorter than those of visible light. The relatively small amount of UV radiation that does reach the ground causes sunburn and can trigger the development of some forms of skin cancer; artificial sources of UV radiation are so harmful to life that they are used to sterilize things like surgical instruments. Without the stratosphere, conditions would be much less comfortable for life on the surface of the Earth.

We enjoy the comfortable conditions underneath the sunshield of the stratosphere because of a complicated series of chemical reactions involving sunlight and molecules of oxygen. The kind of oxygen that we breathe is made up of molecules in which pairs of oxygen atoms are held together by chemical bonds. But it is also possible for oxygen to exist in a tri-atomic form known as ozone – which is poisonous to people. There is a lot of ozone, relatively speaking, in the stratosphere, which is why the stratosphere is sometimes referred to as the ozone layer.

Ozone exists in the stratosphere because ultraviolet radiation from the sun can split apart di-atomic molecules of oxygen, in each case releasing two lone oxygen atoms that can each latch on to another di-atomic molecule to make tri-atomic ozone. If that was the end of the story, all the oxygen in the stratosphere would be converted into ozone, and then solar UV would be able to break through to the atmosphere's lower levels. But ultraviolet radiation can also be absorbed by ozone, knocking off the extra oxygen atom, which can then latch on to yet another di-atomic molecule to make ozone, or instead link up with another spare oxygen atom to make another di-atomic molecule. The overall effect is that ozone is continuously being destroyed and created, at the same rate. And UV radiation is continuously absorbed as part of both processes. So the concentration of ozone in the stratosphere stays the same, rather like the level of the water in a bathtub where the taps are running but there is no plug in the drain to stop the water escaping. (All of this assumes that nothing happens to disrupt the chemical equilibrium of the stratosphere; the kinds of things that might upset that balance are discussed in chapter 10.) From a human point of view, what matters most is that all this activity absorbs dangerous UV radiation, which would otherwise get through to the surface of the Earth.

Surprisingly little ozone is needed to provide this sunshield. In terms of its height, the ozone layer seems to be quite thick – after all, it extends from fifteen to fifty kilometres, a depth of thirty-five kilometres (twenty-two miles), more than twice the thickness, in that sense, of the troposphere. But in terms of its density – the number of molecules bouncing around inside that layer – it is very thin. If all the ozone layer were squeezed to the same pressure as the air we breathe at sea level, with a tempera-ture of fifteen degrees Celsius (sixty degrees Fahrenheit), it would form a layer only three millimetres thick – one-tenth of an inch. That's almost exactly the thickness of a British pound coin, or two US dimes stacked together.

The edge of space

Above the stratopause, there are other, even more tenuous, layers of the atmosphere, which eventually fade away with no distinct upper boundary into empty space. Above an altitude of about fifty kilometres (thirty miles), there is a cooling layer called the mesosphere, which ends at an altitude of around eighty kilo-metres (fifty miles), at the mesopause. Although the density of the mesosphere is low, there are enough atoms and molecules there to provide the friction that burns up small pieces of debris from space, typically the size of sand grains or particles of grit, as shoot-ing stars, or meteors. The temperature of the mesosphere falls off with increasing altitude, to a low of about -120 degrees Celsius, or -185 degrees Fahrenheit, at the mesopause. Above that lies the thermosphere, a layer in which the temperature increases, according to the definition of temperature in terms of the average speed of the molecules, to thousands of degrees Celsius; but there are so few atoms and molecules around that the concept of temperature doesn't really mean anything.

The energy which makes the molecules of the thermosphere move so fast comes from solar X-rays and some types of ultraviolet radiation, so the thermosphere also acts as a shield around the Earth. These forms of radiation are very energetic. The energy absorbed by the atoms and molecules up there is so great that it can knock electrons right out of atoms, leaving positively charged ions behind. For this reason, the thermosphere is also known as the ionosphere. It also 'breathes' in and out, with a twenty-seven-day rhythm related to changes in the amount of energy arriving from the sun, as the sun rotates. Because it is electrically charged, the ionosphere can reflect some frequencies of radio waves, which can be bounced right around the world as a result. Before the advent of communications satellites, this was an important means of global communication, and is still used by amateur radio enthusiasts to talk to people on the other side of the globe.

Even above three hundred kilometres (about two hundred miles), where the International Space Station orbits, there are still billions of atoms of air in every cubic metre. This is only a tiny fraction of the number of molecules in a cubic metre of air at sea level, which is a staggering fifty million billion billion. But it is enough to provide a drag, which lowers the orbit of the Space Station by about thirty metres, or one hundred feet, per day. Fortunately, as the space station's designers anticipated, the station gets a boost to higher altitude when a space vehicle docks with it; this can push it upward by as much as a kilometre – about three thousand feet.

The uppermost regions of the atmosphere, from an altitude of about 500 to 750 kilometres (300 to 450 miles), are sometimes known as the exosphere, because molecules at this height can escape into space. But because there is no clear-cut edge to the atmosphere, the best definition of the limit of the Earth's 'sphere of influence' comes not from trying to find the top of the atmosphere, but from looking at the magnetic field that surrounds us – the magnetic field generated at the heart of our planet, in its core.

Beyond the atmosphere

'Empty' space really isn't all that empty, at least in the vicinity of the Earth. Our planet orbits through a gale of particles seeping past us from the sun, and known as the solar wind. Particles in the solar wind travel at typical speeds of several hundred kilometres per second past the Earth, and at about 1500 kilometres (1000 miles) per second during bursts of activity known as solar storms. The wind extends far out into space beyond the orbits of all the planets, and although it is very tenuous, individual particles in it are very energetic. In addition to the X-rays and ultraviolet radiation mentioned above, and the solar wind particles, we are also bombarded by fast-moving particles from farther out in the universe, called cosmic rays. Cosmic rays and particles of the solar wind could do considerable damage to the atmosphere and to life on the surface of the Earth if they could penetrate the magnetic shield. But because these particles are electrically charged, they are deflected by our home planet's magnetic field and funnelled towards the poles, where they are responsible for the colourful displays of lights in the sky known as auroras.

This magnetic shield around the Earth is called the magnetosphere, but it isn't actually spherical, because it is pushed out of shape by the solar wind. On the side facing the sun, the magnetosphere is squashed in towards the Earth; on the side away from the sun, it stretches out in a long tail. Following the convention for the layers of the atmosphere, the boundary between the planet's magnetic field and the solar wind is called the magneto-pause. Just how far the magnetopause is from the surface of the Earth depends on how strongly the solar wind is blowing. Generally speaking, on the solar side of the Earth it is about ten Earth-radii (some 64,000 kilometres, or 40,000 miles) above the surface of our planet, and on the other side it extends beyond sixty Earth-radii, almost exactly as far as the moon.

The magnetopause marks the true boundary of our home in space.

6

The changing Earth

So far, we've been describing the processes that shape the Earth in impersonal terms, the way, perhaps, an alien scientist might investigate the workings of the planet. But it is, after all, our *home* planet, and even relatively small events on geological scales can have a profound effect on human lives. A few examples bring this home, and demonstrate many of the processes we have described in action.

The Great Lisbon earthquake of 1755

In any roll call of earthquakes, the one that destroyed Lisbon in 1755 stands out. It struck at about 9.40 a.m. on Sunday, 1 November of that year and was one of the most violent earthquakes of historical times. Although it is known as the Lisbon earthquake, because that city was destroyed in the disaster, a large area of the Iberian Peninsula was affected. The region directly disturbed by the shaking of the Earth covered four million square kilometres – 1.5 million square miles. In the city, the earthquake killed an estimated 90,000 out of a population of some 230,000. A further ten thousand were killed across the Mediterranean, in Morocco. And all of this was caused by an event whose centre actually lay far out to sea, in a region now known as the Azores-Gibraltar Fault Zone, part of the boundary between the African and Eurasian plates. It was just five years after the disaster that the English geologist John Michell had suggested that earthquakes are caused by masses of rock shifting their position deep below the surface of the Earth.

With hindsight, you might see this as the seed of the idea that grew to be plate tectonics.

It is impossible to say just how big the Lisbon earthquake was, but it was certainly bigger than the San Francisco earthquake of 1906. Seismologists measure the intensity of earthquakes using a modern version of a scale originally developed in the 1930s at the California Institute of Technology by Charles Richter and his colleague Beno Gutenberg, and known as the Richter scale. The scale measures the energy released in an earthquake based on seismic measurements, and uses what is called a logarithmic scale. This means that an earthquake of magnitude 5 on the Richter scale is a bit more than thirty times more violent than an earthquake of magnitude 4, and an earthquake of magnitude 6 is exactly a thousand times more powerful than an earthquake of magnitude 4, and so on. San Francisco 1906, for a century the popular benchmark, checks in at a magnitude of about 8.3, while the Tōhoku earthquake of 2011, off the coast of Japan, registers at 9. The best estimate is that the great Lisbon earthquake was greater than magnitude 8.8, so it was at least six or seven times the power of the San Francisco earthquake, and possibly as strong as the quake and ensuing tsunami that devastated the north-east coast of Japan. But there were no seismometers around in 1755 to measure how much the Earth shook Lisbon. Instead, we have to rely on graphic eyewitness accounts of the events that November.

One survivor, an Englishman, Thomas Chase, wrote a long letter home to his mother describing what he had experienced. Much later, in February 1813, the letter was published in the *Gentleman's Magazine*, in whose archives the printed version can still be found. When the ground started to shake, Chase was so eager to see what was happening that he climbed to the top of his house to get a good view. He was lucky not to be killed when the house collapsed, flinging him to the ground, unconscious, among the debris. When he came to and dragged himself from the rubble, a Portuguese man who saw the apparition of what seemed to be a dead man walking, 'started back, and crossing himself all over, cried out, as is the custom when much surprised, "Jesus, Mary, and Joseph! Who are you? Where do you come from?"' Chase's account of his state can hardly have comforted his mother:

> My right arm hung down before me motionless, like a great dead weight, the shoulder being out and the bone broken; my stockings were cut to pieces, and my legs covered with wounds, the right ancle [his spelling] swelled to a prodigious size, with a fountain of blood spurting upwards from it: the knee also was much bruised, my left side felt as if beat in, so that I could hardly breathe: all the left side of my face was swelled, the skin beaten off, the blood streaming from it, with a great wound above, and a small one below the eye, and several bruises on my back and head.

The moral? Never run upstairs if you are caught in an earthquake.

Yet, Chase was still in danger. Rescued by friends and put to bed, he now had the terrifying experience of lying helpless while fire swept through the remains of the city. He only survived because his friends carried his bed out into the middle of the great square, which was lit by the burning remains of the king's palace.

For all the drama of Chase's letter, the best indication of the power of the great Lisbon earthquake comes from reports of waves and ripples on ponds far away across Europe, caused by seismic waves travelling outward through the Earth's crust from the centre of the quake. Known as seiches, these waves occur both in the sea (in shallows near the coast) and in landlocked bodies of water. At Portsmouth, England, at ten o'clock in the morning, the forty-gun ship *Gosport* was rocked to and fro as it lay secured in its dock. A few minutes later, in the parish of Cobham, in Surrey, a man watering his horse at a pond fed solely by springs, with no current, was startled – as was his horse – when the water began sloshing about, for no apparent reason. As far away as Prague, between 11 a.m. and noon, spa waters ran muddy, as sediments were stirred by the shaking, a phenomenon nobody could recall ever occurring before.

These seiches were caused by the direct effect of tiny ripples in the surface of the Earth itself, directly shaking the water. But there was also a tsunami after the Lisbon earthquake, a water wave spreading out more slowly from the site of the earthquake. This wave reached Cornwall, on the far west coast of England, several hours later. Fortunately, most of the energy from the wave had been dissipated in the open sea by then, so there was no catastrophic flooding. The wave did, however, reach a height of about three metres, or ten feet, and caused some damage to the coast.

San Francisco 1906

Tsunamis and fire are two of the greatest hazards associated with large earthquakes, and in San Francisco in 1906 it was the fire that caused most of the problems.

Many history books of the great San Francisco earthquake note that it only killed 375 people. But this figure was made up and put out by the authorities, since they thought that if they revealed the real death toll, which was in excess of three thousand, it would frighten people away from the area and hamper attempts to rebuild the city and encourage life (and business) to continue as normal.

In terms of geology, the San Francisco earthquake is interesting because it is the archetypal example of an earthquake occurring where two plate boundaries move past one another, creating what is known as a strike–slip fault. The edges of the plates stick together for a time, until enough strain has built up to overcome the friction between them, then jerk past each other as energy is released, before the whole process repeats itself. It was studies of the 1906 earthquake that led geologists to discover this process, called elastic rebound.

On this particular occasion, 477 kilometres (296 miles) of the San Andreas Fault, running from Cape Mendocino north past San Juan Bautista, ruptured. Along this section, the western side of the fault shifted an average of four metres or thirteen feet northward, relative to the rest of America. In San Francisco, which was not actually at the centre of the earthquake but was the biggest centre of population affected, the first tremors were felt at 5.12 a.m. on Wednesday, 18 April 1906; violent shaking began about twenty seconds later, lasting for up to a minute.

Apart from the loss of life, close to three hundred thousand people out of a population of just over four hundred thousand lost their homes. As much as ninety per cent of this destruction was caused by fires, which raged out of control for four days and nights after the earthquake. The fires were fed by broken gas mains, some of which were wrecked when firefighters used dynamite to blow up houses in a misguided attempt to create firebreaks. Some people set fire to their own homes, in the mistaken belief that they could then claim on their insurance for fire damage even though they were not covered against earthquakes. Even people who were insured against earthquake risks didn't benefit from their prudence, as the insurance companies could not cope with the scale of the disaster; they went bust and failed to pay up. About twenty-five thousand buildings covering nearly five hundred city blocks were destroyed – including the new City Hall, built at a cost of $6 million, some thirty schools, and about eighty churches.

Two years after the quake, many people were still living out of doors in tented 'refugee camps'. The economic consequences of the disaster were dramatic, and are still felt today. Before the earthquake, San Francisco was in effect the capital of the US west, a major economic centre and the port through which trade flowed; it served as the Pacific base for the US Navy. With San Francisco only slowly rebuilding, and fears of another such earthquake, much of this activity was diverted south to Los Angeles, which grew to become the most important city on the US west coast. This may turn out to be highly ironic. Earthquakes like the one of 1906 are indeed bound to recur along the San Andreas Fault; the only questions are when and where. When is harder to answer, but since the southern region of the fault, near Los Angeles, did not rupture last time, logic suggests that the next great California earthquake will actually hit LA.

The Japanese earthquake of 1923

The California earthquake of 1906 provides the archetypal example of an earthquake occurring where two plates rub past one another. The earthquake that hit Japan in 1923, although almost the same size as the San Francisco earthquake at 8.2 on the Richter scale, is, like the disaster of 2011, a classic example of an earthquake where one plate is being pushed under another plate and destroyed. Japan itself is part of an island arc (see page 50), created by this process. We will focus on the 1923 event here, since it is still too soon for the full story of the 2011 quake to be known clearly.

What is known in Japan as the Great Kanto Earthquake struck at precisely 11.58 and 44 seconds on the morning of 1 September 1923. The centre of the event, which lasted for several minutes, was deep beneath Izu Ōshima Island, in Sagami Bay, south-west of Tokyo; but it caused widespread damage throughout the Kanto region, and devastated both Tokyo and the port of Yokohama. One graphic illustration of its power is that a ninety-three-ton statue of the Great Buddha, at Kamakura, shifted by nearly two-thirds of a metre (two feet) during the earthquake. West of Yokohama, the land rose by just over two metres, or 6.5 feet; in other places, it sank by 1.5 metres, just under five feet.

As in San Francisco seventeen years earlier, the worst devastation and loss of life was caused by fire. The earthquake struck just before lunchtime, when food was being cooked over open fires of glowing charcoal and gas stoves in many homes and restaurants. These toppled over and set fire to their surroundings, many of them buildings largely made of wood and paper. The result was a firestorm, fanned by high winds associated with a nearby typhoon. The heat was so intense that many people trying to escape the flames became stuck in melting tarmac and were overwhelmed. The number of known deaths reached one hundred thousand; another forty thousand people simply went missing, and it is assumed that many of them were completely incinerated.

In the mountainous region inland, more destruction was caused by landslides. In one place, the entire village of Nebukawa, its railway station, and a passenger train carrying more than a hundred people were pushed downhill into the sea. Inevitably with an underwater earthquake, a tsunami followed the event, with waves up to ten metres, or thirty feet, high causing further damage and loss of life. Altogether, nearly two million people were left homeless as nearly six hundred thousand dwellings were ruined. The area of total destruction covered sixty per cent of Tokyo and eighty per cent of Yokohama; in Yokohama, one person in twenty died – five per cent of the population.

Such numbers don't really give a feel for the catastrophe, but, as with the Lisbon earthquake of 1755, a foreign visitor has left an eyewitness account that makes things more personal. The Russian writer Petroff Skitaretz, who was moving house when the quake struck, recorded how it felt:

Suddenly, very near the Bluff, I thought I heard the sound of an approaching train. I was surprised, for I knew no train ran near there … From somewhere I heard the roaring as of a wild animal, and a sudden fierce wind came up and bent the branches of the trees like a bow. The sound like an underground train came now from directly under our feet, seemingly some pent up, awful energy seeking escape. The angry roaring increased, an enraged shaking was coming upon us. The ground began to move, groaning and yanking us back and forth with the mad speed and frantic energy of a lunatic. We felt as though we were about to be torn to pieces, and the great earth was trying to shake off everything on it. We seemed to be grains shaken in a sieve. We could not stand and fell in opposite directions, but struggled together against a hedge at the roadside which we grasped.

I looked around. Everything was snapping and cracking, houses and stone walls we had just passed were down … All of these things happened in five or six minutes, and in those few minutes Yokohama and Tokyo were doomed.

The Boxing Day tsunami of 2004

The Great Sumatra-Andaman Earthquake, also known as 'the Boxing Day tsunami', was caused by a magnitude 9.3 earthquake that struck off the west coast of the island of Sumatra, in Indonesia, on 26 December 2004. It was the most powerful earthquake recorded anywhere on Earth in four decades (a 9.5-magnitude earthquake in Chile in 1960 remains the largest earthquake since accurate seismic measurements began). Because of its great magnitude and its underwater location, the Great Sumatra-Andaman Earthquake triggered a tsunami, which swept across the Indian Ocean, devastating the coasts of Indonesia, south Asia, eastern Africa, and Madagascar.

The name *tsunami* comes from a Japanese term meaning 'harbour wave'. Tsunamis are sometimes called tidal waves, though they have nothing to do with the tide, so that expression has fallen out of fashion among scientists. In fact, they don't really have anything in particular to do with harbours, either, but the name tsunami seems to have stuck.

Tsunamis cause so much devastation because a small wave in the ocean can build to a much larger size when it moves into shallow waters – such as when it enters a harbour, or when it moves up a beach. The energy of the wave comes from seismic disturbances associated with sudden vertical shifts in pieces of the Earth's crust at converging or destructive plate boundaries; the Boxing Day earthquake, for example, occurred in a subduction zone. Tsunamis are not triggered at conservative or constructive plate boundaries, where such vertical shifts do not occur; but they can also be produced on a smaller scale by landslides that displace large amounts of water. Where tectonic shifts that trigger tsunamis do occur, a great deal of energy is given up to the ocean above.

On Boxing Day 2004 near Sumatra, at about 8 a.m. local time, 1600 kilometres, or about a thousand miles, of crust dropped down by a distance of fifteen metres (fifty feet) almost exactly north–south along the fault line in the Sunda Trench. Here, the India plate is pushing under a small piece of crust, part of the Eurasian plate that is called the Burma plate. The energy released caused a tsunami, which devastated the immediate region and radiated across the open Indian Ocean. Waves spread out from the entire length of the disturbed crust, carrying the energy equivalent of the explosion of about five megatonnes of TNT – twice the energy of all the explosives used in World War II apart from the two nuclear bombs dropped on Japan. The wave accounted for about a fifth of the surface energy released in the earthquake, which was equivalent to the explosion of twenty-six megatonnes of TNT, about 1500 times the energy released in the Hiroshima bomb. Far more energy was released deep below the surface of the crust during the earthquake. But the open ocean is so big that it can absorb such energy with, literally, scarcely a ripple. It is only when that ripple reaches land that things become more dramatic.

In the open ocean, the wave may have a very long wavelength – even hundreds of kilometres – but a very small height (or amplitude), producing a gentle swell of less than a metre (three feet) on the surface of the sea. These ripples cross the ocean at speeds of hundreds of kilometres per hour – about five hundred kilometres, or three hundred miles, per hour in the case of the Boxing Day Tsunami. All of the energy in the wave has to go somewhere when it reaches the shallows near the shore, and it goes to make a much larger wave crashing over the land. It grows in height and slows to a speed of 'only' about a hundred kilometres per hour, while its wavelength shrinks to less than twenty kilometres. In effect, a shallow but long wave is transformed into a deep but short wave, carrying the same amount of energy. The Boxing Day tsunami reached heights of twenty metres – sixty-five feet – in some places, and in others the water travelled three kilometres, or nearly two miles, inland. Although the devastating effects of the tsunami were felt around the Indian Ocean, the waves even crossed the Pacific, arriving at the western coasts of North and South America in a much dissipated state, but still capable of producing mini-tsunamis with amplitudes of up to forty centimetres, or fifteen inches.

UN figures give the death toll from the disaster, in round terms, as 230,000, of whom 187,000 are known to have died, since their bodies were recovered; the rest are officially recorded as 'missing'. It was the worst tsunami disaster in history, and one of the ten worst earthquake disasters, as well, in terms of death toll. And in terms of energy, it was the fourth most powerful quake on Earth since 1900, and the biggest event in the Indian Ocean region for at least seven hundred years. The economic cost has been estimated at $10 billion.

These almost incomprehensible statistics are put in a more personal context by the fact that eight of those who died were at Rooi Els in South Africa, no less than eight thousand kilometres, or nearly five thousand miles, away from the earthquake. The countries that were severely affected included Indonesia, Sri Lanka, India, Thailand, the Maldives, Sri Lanka, Myanmar, Malaysia, and the Seychelles.

Geological and archaeological evidence shows that there have been at least four major tsunamis, including the one that hit north-east Japan in 2011, affecting the eastern region of the Indian Ocean (and beyond) in the past 1200 years. This reinforces the point that none of the events described in this chapter are unusual as far as our home planet is concerned. They just seem remarkable to us because they are rare on the timescale of a human life. The same is true of the other most visible sign of the power of tectonic forces – volcanoes.

7

Fire on Earth

As we have seen, we know why volcanoes occur, and we know where they occur. But not all volcanoes are the same. Indeed, strictly speaking, no *two* volcanoes are the same. But for convenience, volcanologists classify them in five main categories, even though there are really no distinct boundaries between the categories, and borderline cases are sometimes hard to pigeonhole.

Putting volcanoes in their place

The gentlest volcanic eruptions work in a way that resembles the steady release of lava from the Earth's interior at the spreading ocean ridges. The archetypal example is the Hawaiian volcanic chain, which gives these eruptions their name: Hawaiian volcanoes. These produce very fluid flows of almost pure basalt in a runny form. Because the liquid is so runny, it is easy for any gases associated with it to bubble off, and pressure doesn't build up to the point where the volcano blows its top in a spectacular explosion. It is on the surface of the lakes of fluid magma that fill the craters of such volcanoes that geophysicists can watch slabs of solidified material floating about – a model of the processes of plate tectonics, complete with both constructive and destructive margins and places where 'plates' rub side by side, like miniature San Andreas faults.

Moving up the scale of volcanic violence, the next category are the Strombolian eruptions, which get their name from the volcanic island of Stromboli, which lies between the Italian mainland and Sicily. The basalt that emerges from such volcanoes is thicker and more sticky than the lava that flows from Hawaiian volcanoes, and this allows gas to build up in large bubbles that burst out in minor explosions as often as every few minutes. In these outbursts, lumps of semi-molten lava get thrown up into the air, and sometimes lava breaks out from the crater and flows a little way downhill. But by and large Strombolian volcanoes are more noisy than dangerous, and it is a reasonable rule of thumb that the shorter the interval between explosions, the less dangerous the volcano is. On Stromboli itself, two villages are located within a couple of kilometres of the crater; Mount Etna, in Sicily, is another example. But Strombolian volcanoes are by no means restricted to the Mediterranean region, and Mount Erebus, far away in Antarctica, is another example of the type.

The first really dangerous volcanoes are located roughly in the middle category of the classification scheme, and this is where we find the volcano that gives all volcanoes their name – Vulcano itself, on a small island located to the north of Sicily. (Many volcanic archetypes are located in the Mediterranean, simply because that is where scientists first studied volcanoes; there is no other significance in this.) Vulcano, like other Vulcanian volcanoes, is only active intermittently, but when it does blow its top, it does so more spectacularly than any Strombolian volcano. A Vulcanian eruption can last for months at a time, spewing out blocks of solid material and producing great quantities of ash, carried up into the atmosphere by hot gases rising from the volcano. Although Vulcanian eruptions do often produce flows of molten lava, the ash usually does far more of the damage. The ash spreads over wide areas, and ash falls may continue for a long period.

A Vesuvian eruption, named after Mount Vesuvius, near Naples, is like a more extreme Vulcanian eruption, with a persistent plume of gas and ash rising much higher into the atmosphere. Curiously, because the force of this blast is directed upwards and the ash falls far away, Vesuvian eruptions may cause less damage in the immediate location of the volcano than Vulcanian eruptions do. But this is by no means always the case.

The fifth category of volcanic eruption has two names, or is sometimes subdivided into two essentially indistinguishable sub-categories. A Peléan eruption, named after Mount Pelée in Martinique, in the Caribbean, is a violent outburst that releases hot clouds of gas and dust, which flow down the mountainside and engulf any unfortunate towns and villages that lie in the path. These hot clouds are sometimes known as *nuées ardentes*, the French term for 'glowing clouds', but more commonly as pyroclastic flows. The other name, or sub-category, is a Plinian eruption, named after the Roman Pliny the Elder, who died during the huge eruption of Mount Vesuvius that destroyed Pompeii in AD 79. This eruption produced an enormous vertical column of dust and gas, but it also produced hot clouds of glowing material, which rolled down the mountainside. In this terminology, Plinian eruptions are the biggest volcanic eruptions of all; Peléan eruptions are simply Plinian eruptions, on a smaller scale. The only real difference is that most of the force in a Plinian eruption goes upwards, while most of the blast from a Peléan eruption goes sideways.

In view of their place in popular mythology, it may be a surprise to learn that today there are only about five hundred active volcanoes scattered around the Earth, and probably no more than twenty-five of them will erupt in any one year. But the dramatic impact of even a modest volcanic outburst ensures that these geological actors continue to grab our attention. Geologists have accordingly coined a variety of suitably dramatic names to describe the material thrown out by volcanoes. 'Bombs' are lumps of solid or semi-solid lava that may be bigger than a human being, thrown out hundreds of metres from the centre of activity; if a solid crust surrounds a sticky interior that bursts open when the bomb hits the ground, it is called a 'breadcrust bomb'. Ribbons of lava that twist and solidify as they fly through the air are called, logically enough, ribbon bombs; round lumps are dubbed spherical bombs; long spindly pieces are known as fusiform bombs; and material that makes a splatter on landing is called a cowdung bomb (the children's favourite).

Smaller pieces of ejecta, like grit and pebbles, are called lapilli, and the even smaller bits are simply called ash and dust. The overall name for all the solid material ejected from a volcano is tephra. But it is the smallest pieces of solid material that can do the most damage, as the inhabitants of Pompeii found out.

Vesuvius AD 79

Present-day Vesuvius is the modest remnant of a site of ancient volcanic activity. There, we see a huge crater, part of which can be traced along the semi-circular ridge known today as Monte Somma. The geological activity of the region is a consequence of the collision between the African and Eurasian plates, in which the African plate is being pushed under the Eurasian plate to the north.

Two thousand years ago, the region was geologically quiet and Vesuvius itself had not erupted in living memory. The slopes of the mountain were cultivated with vineyards, which flourished in the fertile soil, and the region was, by the standards of the day, densely populated. In AD 63, there were signs that this period of volcanic inactivity was coming to an end, since Seneca that year recorded a violent earthquake occurring on the mountain and damage to the surrounding towns. Over the next decade and a half, Romans became used to minor earthquakes in the region, regarding them as normal – nothing to worry about. Then, in AD 79, in possibly the most famous volcanic eruption in history, the mountain exploded, burying the nearby cities of Pompeii, Stabiae, and Herculaneum in ash. An eyewitness account of the event has come down to us from Pliny the Younger, the nephew of the elder Pliny. Since then, more modest eruptions have become a regular feature of Vesuvius, changing the size and shape of the mountain dramatically over the centuries, as recorded by many artists. The most recent eruption was in 1944 – and Vesuvius is the only volcano on the European mainland to have erupted since 1900. Naples and its three million inhabitants, just nine kilometres (5.5 miles) from the volcano, can only wait and wonder when, and how big, the next eruption will be.

The younger Pliny, then just seventeen, was staying with his uncle (who was also his adoptive father), Pliny the Elder, at Misenum, thirty-three kilometres (twenty miles) away from the volcano across the Bay of Naples, when the eruption began. He later wrote to Tacitus that:

> On the ninth of the calends of September, about the seventh hour, my mother informed [my uncle] that a cloud appeared of unusual size and shape ... the cloud (the spectators could not distinguish at a distance from what mountain it arose, but it was afterwards found to be Vesuvius) advanced in height; nor can I give you a more just representation of it than the form of a pine-tree, for springing up in a direct line, like a tall trunk, the branches were widely distended ... It sometimes appeared bright, and sometimes black, or spotted, according to the quantities of earth and ashes mixed with it. This was a surprising circumstance, and it deserved, in the opinion of that learned man, to be inquired into more exactly.

The elder Pliny, who was in charge of the Roman fleet at Misenum, then took several boats and set off 'to inquire more closely' into the phenomenon, and help the citizens of Stabiae, about 4.5 kilometres (2.8 miles) from Pompeii. Hampered by the wind, showers of hot cinders, and pumice covering the water, Pliny was forced to stay overnight in Stabiae with a friend. In the morning, the family fled to the boat, heads covered 'with pillows bound with napkins; this was their only defence against the shower of stones. And now, when it was day everywhere else, they were surrounded with darkness, blacker and more dismal than night, which however was sometimes dispersed by several flashes and eruptions from the mountain'.

Even in the boat the family was far from safe. Because of continuing earthquakes that violently disturbed the water, they could not move out into the bay but had to shelter as best they could by the shore until the ash fall ended. There, the elder Pliny died, 'stifled', his nephew surmised, 'by the sulphur and grossness of the air'. But since the rest of the party suffered no such extreme ill effects and there were no marks on the body, it seems more likely that the old and rather corpulent man was overcome by his exertions and suffered a heart attack.

In Misenum, at six o'clock the next morning:

> On the land side a dark and horrible cloud, charged with combustible matter, suddenly broke and shot forth a long trail of fire, in the nature of lightning but in larger flashes ... Not long after the cloud descending covered the whole bay, and we could no longer see the island of Caprea or the promontory of Misenum ...

> We had scarce considered what was to be done, when we were surrounded with darkness, not like the darkness of a cloudy night or when the moon disappears, but such as is in a closed room when all light is excluded. You might then have heard the shrieks of women, the moans of infants, and the outcries of men ...

> A little gleam of light now appeared. It was not daylight, but a forewarning of the approach of some fiery vapour – which, however, discharged itself at a distance from us. Darkness imme-diately succeeded. Then ashes poured down upon us in large

quantities, and heavy, which obliged us frequently to rise and
brush them off, otherwise we had been smothered [or] pressed
to death by their weight …

When the cloud settled and daylight shone on the scene, everything around them was 'covered with ashes as thick as snow'.

The inhabitants of Pompeii were less fortunate. There,
the weight of ashes did smother them to death, entombing
the whole city and preserving the remains until their discovery
in the eighteenth century. It is estimated that about four cubic
kilometres – a cubic mile – of ash and rock was spread over the
area to the south and south-east of Mount Vesuvius by the eruption, which lasted for about nineteen hours.

Krakatau 1883

The other candidate for 'the most famous volcano of all time' has
to be Krakatau, sometimes known by the westernized version of
its name, Krakatoa. The violent explosion of Krakatau in 1883 is
now seen as an almost perfect example of activity associated with
an island arc, in this case a site near Java in Indonesia, on the
western side of the famous 'Ring of Fire', which forms the rim
of the Pacific plate. It is part of the same region of tectonic activity responsible for the Boxing Day tsunami.

After being quiet for a couple of hundred years, Krakatau
showed distinct signs of life on 20 May 1883, erupting with a
noise described as being like distant artillery fire and producing
ash which fell gently on the island of Java, more than 150 kilometres (100 miles) away. Over the next few days, ships passing
through the busy twenty-four-kilometre- (fifteen-mile-) wide
strait between Sumatra and Java reported pumice floating on the
surface of the sea, and a column of dust and smoke rising above
the volcano. On 26 May an expedition to the island found it

covered in white ash, with clouds of material still being spewed high into the air from time to time. The activity was coming from a vent on the side of the mountain, just 120 metres (about 400 feet) from the shore, on a mountain that rose 812 metres (2664 feet) above sea level. With hindsight, it seems likely that the proximity of this active vent to the shoreline contributed significantly to the violence of the explosion that eventually followed.

But the mountain rumbled on alarmingly throughout June, July, and most of August before something even worse than anyone had feared happened. By the afternoon of 26 August 1883, violent eruptions were shaking the mountain, with explosions so loud that they were rattling windows and shaking pictures hanging on the walls of houses in Batavia, more than 160 kilometres – one hundred miles – from Krakatau. There was already now little hope for the few people living on islands near the volcano. But the worst was yet to come.

The explosions which made Krakatau a household name for volcanic aggression occurred on 27 August, at 5.30 a.m., 6.44 a.m., 10.02 a.m., and 10.52 a.m. The third explosion was by far the biggest, according to the distant observers who survived to share their observations. This explosion in particular was so huge that for almost a century it defied explanation. But the best modern supposition – it can never be more than an educated guess – is that the explosions that occurred on 26 August weakened the rock between at least one volcanic magma chamber and the sea, and this collapsed completely on that day, allowing water to flood in and mix with the magma. You might think that water would quench the fire – but where there is so much molten rock and it is so hot, there is no chance that this will happen. Instead, conditions are ripe for what is known as a fuel–coolant interaction, or FCI.

In such an interaction, instead of the coolant quenching the fuel, the fuel is so hot and there is so much of it that as the coolant mixes in to it, the coolant is surrounded and turned into vapour very quickly. Being 'turned into vapour very quickly' is exactly what happens when a conventional explosive is triggered, and the result is the same for an FCI – an explosion. Under the right conditions, a bubble of expanding coolant will fling the surrounding hot fuel, and anything else around it, far and wide. When one cubic centimetre of lava is cooled from 1100 degrees to 100 degrees Celsius (2000 degrees to 212 degrees Fahrenheit), the energy liberated is nearly three hundred joules, meaning that five grams of lava involved in a violent cooling release as much energy as the explosion of one gram of TNT. With tens, or even hundreds, of millions of tonnes of lava, even hotter than 1100 Celsius, present in the magma chambers beneath Krakatau, there is no difficulty comprehending how the mountain blew apart, with a violence estimated as equivalent to the explosion of at least fifty megatonnes of TNT; some estimates suggest it may have been equivalent to a two hundred megatonne bomb. The explosions were so loud that they were heard more than three thousand kilometres away in Australia, and nearly five thousand kilometres away in Mauritius – the latter a distance of three thousand miles.

There were no survivors in the immediate vicinity of Krakatau, but the scale of the disaster that hit Batavia, more than 150 kilometres (100 miles) away, gives a graphic picture of the power of the event. At first, when the mountain began its series of explosions on Sunday, 26 August, the inhabitants of Batavia thought that they were hearing the eruption of another volcano, much closer than Krakatau. It hardly seemed credible that such a distant volcano could literally be rattling their windows. Their alarm increased when the gas supply to much of the island was cut off at about 2 a.m. on 27 August, and although many people did set off to begin the working week, they did so under an overcast that reduced the daytime temperature to as low as seventeen degrees Celsius (sixty-three Fahrenheit) – unprecedented for the location. Fine ash soon began to fall from an increasingly dark sky. As panic set in, thousands of people began to flee from the city, with no clear idea of where to go, but feeling the same urge that the younger Pliny and his companions had – to do *something*. Then, a tsunami, triggered by the mountain's explosion, struck. Batavia city itself, three kilometres, or two miles, from the shore, was flooded to a depth of well over a metre, or about three feet, and the coastal regions of Java and Sumatra were swept by a series of waves, the largest at least thirty metres (one hundred feet) high, that destroyed three hundred towns and villages, killing most of the thirty-six thousand victims of the disaster.

But it wasn't just water that was a killer on this occasion. Hot avalanches of fine ash and volcanic gases spread out from the island of Krakatau as pyroclastic flows. These hot avalanches are bad enough on land, where they can travel at speeds of more than a hundred kilometres, or sixty miles, per hour, engulfing anything in their path. But they can also travel over water. Instead of being quenched, the heat from the pyroclastic flow, typically around 700 degrees Celsius (1300 degrees Fahrenheit), flash-vaporizes the surface of the water, making a layer of steam on which the hot material rides, like a hovercraft riding on a cushion of air. That enabled the hot ash flows from Krakatau to travel up to forty kilometres, or twenty-five miles, across the sea, engulfing ships on the way and causing still more death and destruction when they hit land. Arriving just ahead of the tsunamis, these flows are estimated to have killed about 4500 of the people who died as a result of the 1883 eruption.

When it was safe for boats to venture back into the strait, they found that the geography of this vital link between the Indian Ocean and the South China Sea had been altered. Two-thirds of the island of Krakatau had gone completely, and there were new small islands where none had been before. From a height of 450 metres (1500 feet) above sea level, Krakatau mountain had collapsed into a crater more than 300 metres below sea level; the volcano remains active today. In 1928, a new island dubbed Anak Krakatau ('Child of Krakatau') poked above the surface, and has continued to grow ever since.

It is estimated that the explosion, again, the equivalent of between fifty and two hundred megatonnes of TNT, blasted twenty cubic kilometres (five cubic miles) of rock into fine ash and sent it high into the atmosphere, where it reached heights in excess of twenty-five kilometres, or fifteen miles. There, it spread around the world, producing beautifully coloured sunsets for several years and acting as a sunshield that cooled the entire globe by about half a degree Celsius in the 1880s. Locally, it was so dark that in Djakarta, on Java, people had to use torches and lanterns for the whole day, while in Sumatra they needed artificial light for two days before the skies cleared. The area directly around Krakatau that was affected by falling ash, the fraction of the debris that did not get into the stratosphere, covered four million square kilometres, or 1.5 million square miles, about eight times the area of Spain.

Pelée 1902

Pyroclastic flows played a part in the destruction caused by the eruption of Krakatau in 1883, but only a relatively minor part. When the volcanic Mount Pelée erupted on the French Caribbean island of Martinique in 1902, however, it was pyroclastic flows that did most of the damage. Martinique is part of an island arc, a curved chain of volcanic islands about 850 kilometres, or 525 miles, long, stretching from Puerto Rico to Venezuela. The arc appears where the South American plate is pushing under the piece of crust known as the Caribbean plate, at a rate of a couple of centimetres per year.

Known as 'the Paris of the West Indies', the city of St Pierre was completely destroyed at 7.59 a.m. on 8 May 1902, when Pelée, seven kilometres (four miles) to the north-east, was ripped apart in four deafening explosions, described as being like the cracking of an enormous whip. The explosions had been preceded by activity on the volcano for the previous few days. Nobody had taken this as a sign that anything dramatic was amiss, and no evacuation of the city was attempted, partly because an important political election was imminent. People were also lulled into a false sense of security when they heard that Soufrière, on the nearby island of St Vincent, was erupting; they wrongly thought that this would relieve the pressure on Pelée. In a classic example of bureaucratic complacency, a 'volcano committee' summoned by the governor assessed the situation as follows:

> This phenomenon is normal and commonly observed on all volcanoes around the world. The craters are open so the expansion of the vapours will continue without earthquakes or rock projection. Based on the location of the craters and the valleys leading to the sea, St Pierre is perfectly safe.

But the activity on Pelée did mean that a few observers were out of the city keeping an eye on the volcano, and some of them were lucky enough to survive, as they were not in the path of the pyroclastic flow. These eyewitnesses described a red cloud licking out from the mountain like a gas jet towards St Pierre, with a sound like 'a continuous roar blending with staccato beats like the throbbing of a Gatling gun'. Observers on a passing ship saw the mountainside rip open and a dense black cloud shooting out horizontally.

When it was safe to return to what remained of the city, people saw wood turned to charcoal, iron bars bent into fantastic shapes, and all glass melted. One wrote of a 'desert of desolation, encompassed by appalling silence ... a world beyond the grave'. A French scientist was horrified by the 'pulverized, formless, putrid things which are all that is left of St Pierre'. The damage had mostly been done by hot gas; the flow left behind only a thin layer of white ash, which covered the ruined city.

In the space of a few seconds, the town and its thirty thousand or so inhabitants had been destroyed by a hot, glowing cloud of gas and ash hurtling down the mountain at a speed of several hundred kilometres per hour, with temperatures in excess of 1000 degrees Celsius (about 1800 Fahrenheit), producing devastation which can only be compared to that produced by the explosion of a nuclear bomb. Incredibly, there were two survivors in the city itself. One, Louis-Auguste Cyparis, was a prisoner being held in an underground cell – essentially a dungeon – in the local jail. The other, Léon Compère-Léandre, lived right on the edge of the city, which was only brushed by the pyroclastic flow. Although badly burned, he lived to tell his tale:

> I felt a terrible wind blowing, the earth began to tremble, and the sky suddenly became dark. I turned to go into the house, with great difficulty climbed the three or four steps that separated me from my room, and felt my arms and legs burning, also my body. I dropped upon a table. At this moment four others sought refuge in my room, crying and writhing with pain, although their garments showed no sign of having been touched by flame. At the end of ten minutes one of these, the young Delavaud girl, aged about ten years, fell dead; the others left. I got up and went to another room, where I found the father Delavaud, still clothed and lying on the bed, dead. He was purple and inflated, but the clothing was intact. Crazed and almost overcome, I threw myself on a bed, inert and awaiting death. My senses returned to me in perhaps an hour, when I beheld the roof burning. With sufficient strength left, my legs bleeding and covered with burns, I ran to Fonds-Saint-Denis, six kilometres from Saint-Pierre.

St Helens 1980

The description of 'dense black cloud shooting out horizontally' will have struck a chord with anyone who has seen pictures of the eruption of Mount St Helens, in Washington State. The volcano is located 154 kilometres (ninety-seven miles) south of Seattle and eighty-five kilometres (fifty-three miles) north-east of Portland, Oregon, in the Cascade volcanic arc, which is part of the Pacific Ring of Fire. The trigger for the volcanic activity in this arc is the same set of plate movements that triggered the March 1964 earthquake near Anchorage, Alaska. On a geological timescale, earthquakes and volcanic eruptions are common in the Pacific Northwest of what is now the United States; but the eruption of Mount St Helens that occurred at 8.32 a.m. local time on 18 May 1980, after a hundred years of quiet, was the most lethal and economically disruptive volcanic outburst in the country's history.

Fortunately, in terms of the death toll, that is not saying very much. Only fifty-seven people were killed, thanks to the remote location of the mountain, the fact that it had been showing signs of increasing activity for long enough for people at risk to be evacuated, and because the eruption occurred on a Sunday, when there were no loggers working in the surrounding forest. But 250 homes, forty-seven bridges, twenty-five kilometres (fifteen miles) of railway, and 300 kilometres (175 miles) of highway were destroyed. With about three cubic kilometres of debris – three-quarters of a cubic mile – spewed out in the explosion, the entire top of the mountain disappeared, reducing its height from 2950 metres (9680 feet) to 2550 metres (8370 feet). This left a horseshoe-shaped crater 3000 metres (1800 miles) wide and 800 metres (2600 feet) deep, open at the northern side. Before the eruption, the beautiful symmetry of the peak had often been compared to Japan's Mount Fuji; any similarity was destroyed in the blast. The eruption did not have the global impact of a volcano like Krakatau, however, because most of the ash from the blast came out sideways, instead of being ejected vertically into the stratosphere. About six square kilometres, or 2.5 square miles, of forest containing more than ten million trees was laid flat by the blast, and much of it covered by ash.

In spite of this, as the eruption continued, some of the ash was carried upwards to form a mushroom-shaped cloud, which darkened the sky as it spread downwind in the stratosphere and drifted over eastern Washington State. Travelling at an average speed of one hundred kilometres (sixty miles) per hour, the cloud reached Idaho by noon. The darkness of what became known as 'Black Sunday' eventually covered 57,000 square kilometres – 22,000 square miles – of eastern Washington, northern Idaho, and western Montana. Overall, the eruption lasted for nine hours, but most of the damage occurred in the first seconds, in the initial explosion.

One of the people who died in that blast was David Johnston, a thirty-year-old volcanologist who was monitoring the activity of the mountain from what seemed like a relatively safe position on a nearby ridge. But the sideways explosion sent the pyroclastic flow straight for him at a speed of about three hundred kilometres – nearly two hundred miles – per hour. He was in touch with his colleagues by radio at the time the explosion occurred, so that his last words were recorded for posterity. They were: 'Vancouver! Vancouver! This is it!' No trace of his body was ever found.

This sideways eruption produced the largest avalanche of debris ever recorded, mixing the ash with ice, snow, and water to form mudflows known as lahars. It was these lahars, travelling down the Toutle and Cowlitz rivers at speeds of about seventy-five metres per second – the equivalent of 170 miles per hour – that swept away bridges and did other damage. An estimated three million cubic metres, or 700 cubic miles, of debris was carried all the way into the Columbia River, twenty-seven kilometres (seventeen miles) to the south, by the lahars.

In terms of energy, the explosion of Mount St Helens was equivalent to the blast from twenty-four megatonnes of TNT. It was heard as what people described as 'a thunderous roar' more than three hundred kilometres away. The volcano has remained active since 1980, and there is geological evidence that it has a history of large, explosive eruptions, making it the most active volcano in the entire Cascade Range over the past ten thousand years.

One eyewitness to the eruption was Rowe Findley, a writer for *National Geographic*. In an article for the magazine, he recounted:

> More than fear for personal safety, I felt a growing apprehension for all of us living on a planetary crust so fragilely afloat atop such terrible heats and pressures. Never again, it came to me then and remains with me to this day, would I regain my former complacency about this world we live on.

That seems like a suitable note on which to end this roll call of disasters.

8

Our changing planet

How long has the surface of our planet been moulded by tectonic forces? The oldest rocks found today are almost 4.5 billion years old, so some land must have existed for all that time. Those first pieces of dry land can only have been formed from material scraped up in the process of subduction, or ejected from volcanic hot spots like the one under Hawaii today, to make the first mini-continents. Those mini-continents have grown and parts of them have survived to the present day, while they have been moved around the surface of the Earth by the processes of plate tectonics.

This means that the geography of the Earth has been constantly changing for the past four billion years. There have been times of relative calm, when a particular arrangement of continents has changed only slightly, and periods of more rapid and dramatic change, when supercontinents have been formed or been torn apart; but the underlying theme has always been one of change.

Early times

We know most about the way the geography of the globe has changed for the period since the break-up of Pangaea, which began about 200 million years ago, in the middle of the Mesozoic era of geological time, when dinosaurs walked the Earth. This is because there are abundant fossils to study, and because the rocks themselves have been relatively undisturbed by geophysical processes during that time. The farther back in time we try to probe, the harder it becomes to interpret the evidence. This is partly because of the way rocks have been twisted and deformed, become molten and been reworked during repeated bouts of continental drift, collisions between continents, and ripping apart of continents along great rift valleys. But it is also because fossils are harder to come by in very old strata. There are very few fossils at all to be found more than about 600 million years old, in the great stretch of geological time known as the Precambrian, which spans roughly eighty-five per cent of Earth history. This is because it was only in the Cambrian period, beginning about 550 million years ago, that creatures with shells and bones evolved. Shells and bones make good fossils; the soft-bodied life forms of the Precambrian did not. The very fact that so much of Earth history is lumped together as the Precambrian tells you just how important fossils are to geologists and geophysicists.

Nevertheless, it is possible to infer that, for example, by about three billion years ago roughly half of the land that would become present-day North America had formed. Since then, as they have been moved around the globe, continents have grown slowly, by accretion, as more continental crust has been added around their edges.

In Archaean times (roughly between four billion and 2.5 billion years ago), there seem to have been several small continents with much thinner crust than the present-day continents, being carried about more rapidly than the present continents on the surface of a much hotter Earth which was experiencing more vigorous convection. Things began to change in the Proterozoic, from about 2.5 billion years ago to about 545 million years ago, when there were larger continents (some formed by the merging of the original small continents), which may from time to time have joined together to make a single supercontinent.

Such evidence as there is suggests that about 1.5 billion years ago most of the landmasses of the Earth were joined in a single supercontinent, a proto-Pangaea, but that this was not a particularly stable configuration. There were certainly intervals when parts of what later became Gondwana and North America split apart before coming back together again, and other gaps may have opened and closed together again over a span of hundreds of millions of years. A bigger split occurred some time in the late Proterozoic, producing five distinct continents. Four of these were recognizable early versions of North America, Europe, Siberia, and China, although they did not occupy the same positions on the globe then as they do now. The fifth was the southern continent of Gondwana. The four northern continents did not have an independent existence for long, as geological time is reckoned. The dance of the continents soon brought the land that now comprises Scotland, Greenland, North America, Scandinavia, and Russia together in a continent sometimes known as Laurussia, which lay just north of the equator some 320 million years ago. To the south, extending as far as the South Pole, lay Gondwana, while to the north lay the cores of present-day Siberia and China. By 250 million years ago, all the major landmasses had come together once again to form Pangaea, which covered two-fifths of the Earth's surface in a single continuous continent.

Pangaea had an irregular outline whose principal feature was an enormous triangular notch (like a V turned on its side, <) halfway down its eastern side. This giant bay formed what is known today as the Tethys Ocean. Beyond this bay was a world ocean, dubbed Panthalassa.

Roughly speaking, the region of Pangaea north of the Tethys Ocean contained the cores of the northern continents already mentioned, covering an area of about eighty million square kilometres, or about thirty million square miles. Gondwana, south of the Tethys Ocean, was roughly the same size and included what would later become South America, Africa, India, Australia, and Antarctica. The interior of Pangaea, far from the ocean, was a vast region of desert, dried-out salt lakes, and mountain ranges; near the coast there were swamps and forests, in which reptiles and early dinosaurs roamed. But life did not begin to dominate the land until Pangaea broke up, and rain-bearing winds from the sea could reach the interiors of the smaller continents that subsequently were produced.

It was the break-up of Pangaea, beginning about 200 million years ago, that produced the world as we know it today; and it is this phase of continental drift that is most clearly understood from fossils and other geological evidence.

Two million centuries

Compared with some of the numbers we have been bandying about, 200 million years sounds like a blink of the eye of geological time, and in a sense it is. But just how long a time it really is in human terms is brought home to us if we remember that this corresponds to two million centuries, compared with the twenty centuries of the entire Christian era to date. One of the most popular ways to make changes taking place over such vast stretches of time intelligible in human terms is to imagine the whole span compressed into a single day, or a single hour. Following that hallowed tradition, let's look at the past two million centuries of Earth history in terms of sixty minutes ticking by.

As the clock starts ticking, the Tethys Ocean lies almost on the equator, with about as much land in the northern hemisphere as in the southern hemisphere. North America, South America, and Africa are joined, in the northern part of Pangaea, in a three-way junction about where Ascension Island, in the South Atlantic, is today. Present-day New York is on the equator, while the rocks that will become Japan are in the Arctic. In the southern part of Pangaea, the land that will become India and Australia is far to the south, linked to what will become Antarctica, which is already situated across the South Pole.

As the clock begins to tick – let's say, at high noon – the weak link in the V (or <) where northern and southern Pangaea are joined alongside the Tethys Ocean begins to break apart. This zone has never been strongly welded together, and upwelling of basalt from the hot interior of the Earth has stretched the surface and made cracks and fissures like the present-day Great Rift Valley, in Africa.

Just as in the Great Rift Valley, the floor of these fissures has been sinking as the sides have moved apart, and at noon the waters break in from Panthalassa to create a new narrow sea, like the modern Red Sea, separating North America from South America. At first this new sea is closed at one end, in the east. Just like the Red Sea today, where Arabia and Africa touch in the north, the land that will become Spain is still joined, but just barely, to the land that will become Africa. As Laurasia, made up of modern North America and Eurasia, begins to move north, this place of connection acts as a kind of hinge, so that the whole Laurasian plate rotates clockwise, as viewed from above. This has the effect of opening up the rift in the west, widening the gap between North America and South America, and forming what would later become the Atlantic Ocean. At the same time, in the east this rotation begins to close the Tethys Ocean, which will eventually shrink to become a remnant of itself, known today as the Mediterranean. The floor of the Tethys Ocean is steadily consumed in a trench lying from present-day Gibraltar to modern Borneo.

While all this is going on, in the southern hemisphere Gondwana is splitting into two pieces, with South America and Africa (still joined together at this time) on one side of a new rift, and the block of continental crust made up initially of Antarctica, Australia, India, and Madagascar on the other. All of this activity, the complete break-up of Pangaea, is complete by twenty minutes past twelve on our geological clock (135 million years ago in real time), and by then India has already split off on its own journey northward. Over the next ten minutes (or thirty-three million years), the young North Atlantic Ocean grows to a width of about a thousand kilometres (six hundred miles) in the south, while in the north the crack in the continental crust that made the young ocean extends farther and splits Greenland away from Canada. At this time, what will become the east coast of the United States is aligned almost at right angles to its modern position, running roughly east–west at a latitude of twenty-five degrees north; evidence for this comes from the remains of coral reefs found today all along the eastern edge of the continental shelf of North America. Meanwhile, on the other side of the Atlantic, the grinding of the African plate against the Eurasian plate twists what will become the Iberian Peninsula in an anti-clockwise direction, opening up the Bay of Biscay.

One hundred million years after the ocean flooded in to the rift between Laurasia and Gondwana, at 12.30 on our clock, Africa and South America begin to split apart and the South Atlantic Ocean is born. Over the next ten minutes (thirty-three million years) the geography of the globe takes on a recognizably modern appearance. The rift between South America and Africa joins up with the rift in the north to become the Mid-Atlantic Ridge, while at its northern end the rift forks around Greenland, splitting it away from Eurasia as well as from North America. North America begins to rotate anti-clockwise into its present orientation. No longer attached to South America, Africa shifts north and east, finally closing off the remnant of the Tethys Ocean but breaking the link between Africa and Eurasia across what is now the Strait of Gibraltar. But the bumping of Africa against Europe at the western end of the Mediterranean will repeatedly open and close this gap, and force up the mountains of the Pyrenees to the north.

As North and South America move steadily westward, more or less in their present orientations, but still with a gap between the two continents, they are crumpled up along their western edges to form the Rockies and the Andes. Farther south, New Zealand splits off from Australia and Antarctica, and in the final stage of the break-up of Gondwana, Australia splits off from Antarctica. By this time, India has already split off from Australia, and is about to be involved in one of the most spectacular geophysical events of the past fifty million years.

With just over a quarter of an hour to go on our clock, about fifty million years ago, India is being carried north across the equator, ploughing into the sediments laid down on the floor of the ancient Tethys Ocean and piling them up along its leading edge. Just after crossing the equator, like a car crumpling as it crashes head-on into a solid wall, India rams into Eurasia with such momentum that the northern edge of India is shortened by two thousand kilometres, or 1200 miles, as the crust is squashed into a layer fifty-five kilometres (thirty-five miles) thick beneath what becomes the Himalayas and seventy kilometres (forty-five miles) thick beneath what becomes Tibet. Fossils of marine creatures from the time of Pangaea, laid down in the Tethys Ocean, are forced upwards with the rocks, to be found today far inland on the peaks of the highest mountains in the world.

In the west of the combined Africa-Eurasia landmass, the grinding of Africa against Europe is producing mountain chains all the way from the Himalayas through the Caucasus and the Alps to the Atlas mountains of Africa. One sliver of Africa, Italy, is forced right across the Mediterranean and into Europe, pushing the Alps higher in an echo of the uplifting of the Himalayas when India rammed into Eurasia.

Just seven minutes before one o'clock, about twenty-three million years ago, the waters flowing eastward from the everrising Andes carve their way to the Atlantic, forming the Amazonian river system. The crack in the continental crust forming the Red Sea–Gulf of Aden system also begins to open about now, just as Africa becomes firmly linked to Eurasia at the eastern end of the Mediterranean. Now, the only link between the

Mediterranean and the world's oceans is at the Gibraltar Strait, and the next time it closes, a little after five to one (about fourteen million years ago) the Mediterranean dries up, forming a thick layer of salt. The waters of the Atlantic don't break through to the Mediterranean again for a couple of minutes (until about six million years ago). When they do, about a twentieth of all the world's ocean water pours over the shelf of rock between southern Spain and northern Africa in a waterfall that lasts for roughly a hundred years – which corresponds to one five-hundredth of a second on our clock. It takes so much water to fill the Mediterranean that the level of the oceans around the world drops by about twelve metres, or forty feet.

Because the floor of the Mediterranean itself has been reworked repeatedly by the encounters between Africa and Europe, it probably contains little or no trace of its origins in the Tethys Ocean. The best place to look for a 'fossil' remnant of the Tethys Ocean is farther to the east, in the Black Sea, between Turkey and Ukraine.

Almost with the last tick of our clock at one minute to one, 3.5 million years ago, the Isthmus of Panama is formed by geological activity as South America closes in on North America, linking the two Americas and putting the last piece of the modern picture of the globe in place. But tectonic activity has not stopped, as the evidence of the previous two chapters testifies, and the plates are still moving. So what will the world look like 100 million, or 200 million, years from now?

Future worlds

Over the past 500 million years, the geography of our changing planet has reflected the way in which Pangaea was put together out of smaller pieces of continental crust, then torn apart to make the present arrangement of the continents. Pangaea itself existed for only a relatively brief moment of geological time – no more than about 100 million years. But, according to the best guesses we can make about the future drift of the continents, in about 250 million years from now, a new Pangaea may have formed.

Those best guesses are the work of Christopher Scotese, a geophysicist working in the 1980s at the University of Chicago and more recently at the University of Texas in Arlington, and were presented to the public in his book *Continents in Collision*. Just as with our understanding of past tectonic movements, his projections are likely to be more accurate for dates that are closer to us in time, and more speculative for those that are more distant. He has based them on the present patterns of tectonic activity around the Earth, together with what amount to educated guesses about where new rift valleys and subduction zones are likely to form. They should not be taken as a literal forecast of things to come, but as an indication of what might be.

One hundred million years from now, if the plates keep moving as they are today, the Atlantic will have grown to become the world's biggest ocean, while the Pacific will have shrunk from its former glory. The Mediterranean Sea will have disappeared as Africa forces its way northward, with new mountain chains building up across what is now Europe and all the way east to the Himalayas. Water will have flooded into the widening African Great Rift Valley, separating a sliver of Africa from the main body of the continent, and the part of California west of the San Andreas Fault will have been carried north to Alaska. Australia, relentlessly grinding north towards Asia, will have squashed Borneo and the other islands in between into new mountain ranges. (Some geophysicists suggest that Australia may also rotate, eventually colliding with China, but Scotese does not go that far.) South America will have pushed up towards Florida, beginning to close the Caribbean. All this is reasonably certain. Slightly more speculatively, Scotese suggests that new subduction zones, complete with their associated arcs of volcanic islands, will become active along the eastern coasts of both North and South America, and around Australia–New Zealand.

As a result of the activity along these new subduction zones, over the following fifty million years both the Indian and the Atlantic Oceans will shrink. According to Scotese's scenario, at the same time new rifts will open up in the Pacific sea floor, which will once again begin to grow. One effect of this will be to tear apart the link between Alaska and Siberia, perhaps with a fragment of the former Siberia being carried eastward with Alaska. By now, the fragment of California that had travelled north will have lost its identity and become part of a crumpled range of new mountains. The African rift has become a baby ocean, and the subduction activity south of Australia (which has itself merged with New Zealand) has closed the gap between Australia and Antarctica, with Antarctica having been carried north on the tectonic conveyor belt to become welded to Australia.

Scotese's final projection is of the world as it might appear in 250 million years' time. The Atlantic and Indian Oceans have completely closed, and the African Rift Ocean has become an inland sea. The collision of continents has created a new Pangaea, stretching almost from the South Pole to the North Pole. Traces of what were coastlines can still be seen in the form of mountain ranges created in the collisions between continents.

This is just one possible future world, and one thing we can be sure of is that the real world will not develop exactly like this. But we can also be sure that if and when a new Pangaea does form, it will not last long before tectonic forces tear it apart, in a renewed dance of the continents. This is almost all we can say about the changing geography of our planet. But there is one intriguing, if controversial, idea that deserves at least a passing mention before we look at the way the geography of the globe affects the climate of our planet, and may have influenced the evolution of humankind.

An intriguing hypothesis

When geophysicists try to reconstruct past arrangements of the continents, by drawing maps with the help of computers of how the world used to be, they are in effect working with a spherical jigsaw puzzle, sliding the pieces of crust around on the surface of a globe to make them all fit. You cannot do the same sort of thing so successfully with a flat map of the world, because any flat map distorts the relationships between continents on our round planet. Any flat map of the world is actually a 'projection' of a more or less spherical surface onto a flat piece of paper, and although there are different projections that are useful for different purposes, none of them is perfect. For example, one projection may be good at representing the shapes of continents, but gets their relative sizes wrong, while another is good at getting the relative sizes right, but to do so it must distort the shapes of the continents.

It's easy to see the fundamental problem by peeling an orange very carefully so as to keep the skin in as few pieces as possible. If you lay the pieces of orange peel out flat on a table, you always leave gaps, and cannot get the pieces to join up in a complete disc. There are always gaps somewhere in the pattern, even though on a round surface the pieces fit together neatly to make a smooth and continuous whole. So reconstructions of Pangaea, for example, have to be made for continents drifting about on the surface of a sphere.

Most reconstructions of Pangaea are made for a sphere the same size as the Earth is today. That's natural enough. These projections, like Teddy Bullard's matching up of Africa and the Americas, are nearly perfect, and certainly better than similar reconstructions made by assuming the Earth is flat. But there are still relatively small gaps in the reconstructed pattern, where the pieces of the continental crust don't quite fit together. The usual explanation for this is that our geological records are imperfect, and there may have been small changes in the edges of the continents that we are not aware of. There's nothing controversial in this.

But here is where the intrigue begins – as well as the controversy. While building on earlier work which pre-dated the modern theory of plate tectonics, Hugh Owen at the British Museum, in London, discovered that you can get the fit to be almost perfect if the reconstruction is made for a globe with eighty per cent of the Earth's present diameter. Taken at face value, this means that, since twenty is a quarter of eighty, the Earth has expanded by a quarter of its size in the early Jurassic, over the past 200 million years or so. It also would mean that the Tethys Ocean never existed – that whole V-shaped feature would simply be interpreted as a gap in the reconstruction, caused by using the wrong diameter for the Earth.

There is other evidence to support Owen's hypothesis of expansion. For example, magnetic evidence tells us that all the continents except Antarctica are drifting northwards, which ought to mean that the Arctic Ocean is getting smaller. Yet, the Arctic zone seems to be getting bigger. The two sets of data can be reconciled, of course, if the Earth is expanding.

Even if Owen's suggestion is correct, however, this does not mean that all of continental drift and plate tectonics can be explained in terms of an expanding Earth. There is overwhelming evidence for seafloor spreading, subduction, and all of the other activity we have described, and, for example, no direct evidence that the Earth has expanded enough to create the huge area covered by the Pacific Ocean. Owen suggested that something unusual may have happened to the planet about 200 million years ago, making it swell up and cracking the crust, triggering the break-up of Pangaea. The kind of process that might produce such an effect would be a change in the nature of the core of the Earth, similar to the kind of change that happens when water turns to ice, involving a transition from a high-density state to a lower-density one. A securely fastened bottle full of water left in a deep freeze will be cracked apart as the water expands on freezing, because of just such a change, known as a phase transition. A phase transition in the core, associated with the crystallization of the liquid outer core onto the solid inner core, might increase the volume of the core, and perhaps release extra heat, inflating the whole planet and pushing the skin of the Earth outward as it expands. But this is pure speculation.

There are many problems with the expanding Earth hypothesis. Direct evidence using astronomical measurements and data from satellites is now accurate enough to tell us that the Earth is not expanding much, if at all, today. But of course, that does not mean its size has never increased! Another problem is that if the Earth has expanded as much as Owen suggested, the strength of gravity at the surface would have got a lot lower, because the surface would be farther out from the centre of the planet. And there is no geological evidence for that, although it is very difficult to interpret what evidence there is. The expanding Earth would also spin more slowly, like a spinning ice-skater who stretches out her arms – again, no evidence for that, either. And yet, Owen's reconstructions of Pangaea and its components undeniably work better for a smaller Earth.

The expanding Earth hypothesis is probably wrong, but the evidence of these reconstructions ought to encourage geophysicists to think of other explanations for the gaps that are left when the reconstructions are made for a sphere the size of the Earth today. After all, however you look at it, the evidence is telling us that something we don't yet understand is going on. Since continental drift itself has been a respectable idea for barely half a century, it would be foolish to imagine that we already have all the answers – and remember, it was maps like Owen's that first drew attention to the 'impossible' hypothesis (as it then was) of continental drift.

9

Our changing climate

Geologists reconstruct the history of past climate from the traces left in rocks. In the most extreme example of this, the movement of ice sheets across the surface of the Earth during ice ages has gouged out scars which are still visible hundreds of millions of years after the ice has gone. By dating different rock strata in the ways we have described, and looking for traces of ice movements, it is possible to determine, to a reasonable accuracy, when those particular pieces of continental crust were covered by ice.

But that is only half the story. It is just as important to determine *where* those pieces of crust were at the time they were covered by ice. As ever, all of this becomes more difficult the farther back in time we try to probe. We have a good idea of at least the broad pattern of ice ages and warmer intervals over the past 600 million to 700 million years, and a rather more vague idea about how the climate of our planet varied before then. But we have a very good idea of the way our climate has changed over the past five million years or so, the time during which our own species, *Homo sapiens*, has emerged.

Present patterns

Over the longest intervals for which we can say anything with confidence about the changing climate of our planet, the past 700 million years or so, the 'normal' climate of the Earth seems to have been warm and wet. At intervals of very roughly

250 million years, a change occurs that brings with it a colder climate regime, sometimes called an ice epoch, which lasts for millions of years. Within an ice epoch, glaciers and ice sheets ebb and flow many times, sometimes spreading out in a full ice age, and sometimes retreating part of the way towards the poles in a slightly warmer, and much shorter, interval known as an interglacial. We are living in an interglacial today.

The slender evidence that ice epochs recur at regular intervals has encouraged speculation that they might be caused by some outside agency. This could be a rhythmic (but unexplained) change in the output of the sun itself, or an effect linked to the motion of the entire solar system in its orbit around the centre of the Milky Way galaxy, which itself takes a couple of hundred million years. But it is more likely that the apparent rhythm is just a coincidence, and that the root cause of ice epochs can be found in the changing geography of our planet.

Evidence for this comes from looking at the pattern of ice on Earth today. Apart from on high mountains, we find ice sheets and glaciers near the poles. This seems natural to us, because it is what we are used to, but it has not been natural for most of Earth history. Near the poles, the sun sits low on the horizon or is completely absent for much of the year, and less heat reaches each square metre of ground than at lower latitudes. If the warm waters of the ocean could penetrate to the poles, they would warm them and prevent ice forming. But in the south, the continent of Antarctica sits squarely over the pole, preventing this and providing a platform on which great ice sheets can build up. In the north, the situation is slightly more subtle. True, there is an ocean over the pole; but the Arctic Ocean is almost completely landlocked, and it is difficult for warm water to penetrate it, so a thin covering of ice has formed over its surface, while ice sheets have grown over the Arctic land surface of Greenland.

When we find, as geologists do, that there are the fossilized remains of tropical forests in Antarctica, we do not assume that this means that the South Pole was once covered in tropical jungle, but that the land that is now Antarctica used to be near the equator. Similarly, when we find evidence of glaciation in Brazil, we don't assume that ice covered the entire globe at that time, but that the land that is now Brazil used to be part of a polar continent. So continental drift alone can explain the pattern of major ice epochs.

Ancient history

Before about 500 million years ago, at a time when all of the continents were assembled in an early version of the single super-continent, Pangaea, there was a great ice epoch. At that time, the supercontinent overlapped one of the poles, allowing ice sheets to build up just as they have over Antarctica today. There was no land over the other pole, so there is no evidence in the sedimen-tary record of what the climate there was like at that time. But it is a reasonable guess that since warm water could still penetrate to the region, it was warm and wet in that hemisphere of the Earth, even though the other hemisphere was in the grip of ice.

When this version of Pangaea broke up, its four main pieces drifted off in different directions and warm water could get to both poles again. This was a time, the Devonian period of geo-logical time, which ended about 350 million years ago, when life flourished both in the sea and on land. (Life first appeared on land nearly 500 million years ago, in what is called the Ordovician period.) During the Devonian, both plants and animals diversi-fied into many different forms. But things became harsher for life as the land fragments re-assembled to form Pangaea proper. During this assembly process, the landmass made up of present-day Africa, Antarctica, Australia, and South America drifted across the South Pole, causing a glaciation, which has left its scars on all four of these modern continents. Even after Pangaea was assembled, the southern part of the supercontinent extended across the South Pole, so the glaciation continued in the south even though a great deal of the landmass extended up across the tropics. The North Polar sea, though, was warm and ice-free.

After about fifty million years of this one-sided ice epoch, Pangaea began to break up again and drifted away from the South Pole. The whole Earth became warm, with snow and ice confined to the tops of high mountains. Once again, life flourished, and now it spread across the land. It had suffered a major setback in a wave of extinctions that struck at the end of the Permian period, some 250 million years before, when the southern continents had drifted over the pole yet again – but this also opened the way for new species to evolve. Dinosaurs roamed the globe for nearly 150 million years, beginning at the end of the Triassic period, about 205 million years ago. During this time, the separated continents were edged by shallow seas in which life thrived, and warm, rain-bearing winds could easily penetrate into the interiors of the relatively small continents. But about fifty-five million years ago, as the geography of our planet began to take on the appearance familiar to us, the world began to cool and the climate also began to change into the pattern we know today.

As Antarctica separated from Australia, the former slowly crept into its present position over the South Pole. Meanwhile, to the north, as the Atlantic opened, the great ocean on the other side of the world began to be squeezed out of existence, forming the smaller Pacific Ocean as we know it today. At the same time, the northern continents were edging farther north, restricting the access of warm water to the Arctic basin more and more as time passed.

By about ten million years ago, the present ice epoch was firmly established. Glaciers had already appeared in Alaska and in other places around the Arctic Ocean, and the ice sheet over Antarctica was about half as big as it is today. By about five million years ago, the Antarctic ice sheet was even bigger than it is today, and by about three million years ago great ice sheets had appeared over the continental masses of the north, pushing down into Europe and North America. The pattern of geography and climate that we think of as normal had become established.

Ever since then, the ice sheets in the north have ebbed and flowed in a complex pattern, which follows a variety of rhythms roughly 100,000, 42,000, and 22,000 years long. Very approximately, the combined effect of these rhythms is that intervals of full ice age conditions of about one hundred thousand years are separated by intervals of interglacial conditions, like the conditions on Earth today, of about ten thousand years. Humankind has evolved during this ice epoch, and human civilization has developed during the present interglacial, so these rhythms are of much more than academic interest. Indeed, as we explain in the next chapter, they are the reason why we are here.

But why has the climate followed this pattern for millions of years? The timing of these rhythms is very well understood, and depends on an understanding of the nature of the Earth as a planet orbiting a star, the sun. But just why the ice age–interglacial pulse should beat so strongly is slightly more of a mystery.

Victorian superseasons

The realization that the climate of the Earth changes on a geological timescale came almost as soon as scientists realized that geology required a much longer timescale than the timescales of human history, or of biblical chronologies. James Hutton, the Scot who was one of the founding fathers of the science of geology, visited the Jura mountains of France and Switzerland in the eighteenth century, and noticed the jumbled heaps of rock that lie in many of the valleys far below the altitude of the glaciers today. These are just like the piles of rubble, called terminal moraines, that accumulate at the ends of glaciers, where the melting ice dumps debris that it has carried down from the heights. Hutton grasped that he was seeing evidence that the glaciers used to extend farther down into the valleys, back when Europe was much colder and the ice spread to lower altitudes and to lower latitudes. He published this conclusion in the 1790s, but nobody took much notice at the time. In the first decades of the nineteenth century, most geologists preferred the argument that boulders found far away from the strata to which they belonged had been carried there by water, not by ice; and to those of a religious persuasion this was seen as evidence for the biblical Flood.

It was a Swiss expert in fossil fishes, Louis Agassiz, who eventually took up the ice age idea and promoted it vigorously. In 1837 Agassiz, although only thirty, was named president of the Swiss Society of Natural Sciences, and he startled the learned members of the society by choosing for his presidential address not the lecture on fossil fishes that they expected, but a dramatic endorsement of the idea of climatic change, in which he introduced the term 'ice age' (*Eiszeit*). He made few converts at the time, but this only encouraged him to make more efforts to find evidence for the ice sheets that had once covered much of Europe, and to keep promoting the ice age idea to other geologists, including Charles Lyell. By 1840, Lyell was convinced, and in the same year Agassiz himself published a book, *Études sur les glaciers* (*Study on Glaciers*), which helped to convince others.

From then on, the idea that at least one ice age had existed was deemed to be respectable science, even if it was not yet endorsed by all geologists. And as the years passed and more evidence accumulated, it became clear that there had been not one, but several ice ages. Scientists in Victorian times took on the view that the Earth experiences a pattern of superseasons, the kind of ice age rhythms the Earth has experienced over the past few million years. What could be causing them?

One clue to the superseasons comes from understanding the cause of the seasons themselves. We have seasons on Earth because the planet is tilted in its orbit. Imagine a line from the sun to the Earth. This line, instead of making a right angle with a line through the North and South poles of the Earth, makes an angle of 66.5 degrees – in other words, the Earth's tilt is 23.5 degrees out of the vertical.

Over the course of a year, the North Pole always 'points' in the same direction – as it happens, towards a particular star, known as the Pole Star. But as the Earth orbits round the sun, this means that in one part of the orbit the northern hemisphere leans towards the sun, while on the other side of the orbit the northern hemisphere leans away from the sun – just as if the sun had gone around behind the Earth. In between, there are two days in each year, on opposite sides of the orbit, when the sun is exactly on one side of the Earth, measured relative to a line through the poles. These days are called the equinoxes, which occur in the spring and the autumn.

When the hemisphere is tilted towards the sun, people living in that hemisphere see the sun rise high in the sky, and the pole itself never sees the sun set. It is summer. When the hemisphere is tilted away from the sun, people living in that hemisphere see the sun rising only a little way in the sky even at noon, and the pole itself is dark twenty-four hours a day. It is winter. Because the sun rises higher in the sky in summer, there are more hours of sunlight, and this is one reason why summer is warmer than winter. But when the sun is high in the sky, a patch of ground or sea gets more concentrated sunlight than the same patch does when the sun is low in the sky. This is why the sun feels hotter at noon than in the morning or in the evening.

Now, imagine a 'beam' of sunlight one metre square arriving at an angle on the surface of the Earth. The more shallow the angle, the more the beam gets spread out over the planet's surface. At high noon in summer at the latitude of New York City (which is roughly at forty degrees North), each square metre (about ten square feet) of radiation from the sun spreads out over 1.25 square metres (about 13.5 square feet) of the Earth's surface; but at high noon in winter, the same square metre of solar radiation is spread over 2.5 square metres (about twenty-seven square feet) of the metropolis. That, and the fact that there are fewer hours of daylight in winter than in summer, is why winter is colder than summer.

The sun is at its highest in the sky in the northern hemisphere on 22 June, the summer solstice; it is at its lowest noon altitude on 22 December, the winter solstice. All of this pattern of the seasons is reversed in the southern hemisphere. At the equinoxes, on 21 March and 23 September, everywhere on Earth the sun rises exactly to the east and sets exactly to the west, and there are twelve hours of daylight and twelve hours of night.

Because the Earth's orbit is not quite circular, we are at our closest to the sun (perihelion) on 3 January, near the time of the northern hemisphere winter solstice; we are at our farthest from the sun (aphelion) on 3 July. This is not the cause of the seasons, but it means that northern hemisphere winters are a tiny bit less cold than they would be if the Earth's orbit were circular, and southern hemisphere summers are a tiny bit cooler than they would be if that were the case. And it turns out that this non-circularity of the Earth's orbit *does* play a part in the superseasons called ice ages.

Astronomy and ice ages

Several people groped towards an understanding of the superseasons in the middle of the nineteenth century, but the person who made the breakthrough was James Croll, who had been born in 1821. Croll came from a humble Scottish background, and his formal education came to an end at the age of thirteen; but he devoured books about science and educated himself as best he could while making a living (just) in a succession of short-lived jobs. His big break came in 1859 (the year Charles Darwin's book *On the Origin of Species* was published), when he found work as a janitor at the Andersonian College and Museum, in Glasgow. At the age of thirty-eight, Croll at last had a place where, in between his janitoring, he could toil quietly at his own ideas, and he had access to a first-class library.

He made such good use of the opportunity that within a couple of years he was publishing scientific papers on electricity and other topics. In 1864 he published his first paper on ice ages. As his standing in science grew, in 1867 he got a job with the Geological Survey of Scotland. It was officially a clerical position, but that was a technicality used by the survey's director, Archibald Geikie, to get around the fact that Croll was officially unqualified for an academic post. Croll now had an academic home where he could work on his research full time. In 1876 Croll was elected as a fellow of the Royal Society, soon after receiving an honorary degree from the University of St Andrews. And his 'big idea' for which he won acclaim was an astronomical theory of ice ages.

Croll studied the way in which the Earth's orbit around the sun changes over time. Sometimes the orbit is more elliptical, and sometimes it is very nearly circular, a well-known pattern linked with the gravitational influence that the planets in the solar system have on each other. It is simple to calculate in principle, but this was very tedious to do in the days before astronomers had electronic computers to do the number-crunching for them. The person who had crunched the numbers by hand, literally with pencil (or pen) and paper, was the French mathematician Urbain Le Verrier, whose work Croll used as the basis for his ice age studies.

It turns out that the eccentricity of the Earth's orbit (which measures its deviation from circularity) varies over a cycle roughly one hundred thousand years long. Today, the eccentricity of the orbit is about one per cent, but at its greatest the eccentricity is about six per cent. Because the Earth was in a high-eccentricity state one hundred thousand years ago, while it has been in a low eccentricity state for the past ten thousand years, and because it is warmer now than in the past, Croll guessed that something to do with high eccentricity caused ice ages.

One important feature of these changes in the Earth's orbit is that they do not alter the total amount of heat received by the Earth from the sun over the course of a year. In a high-eccentricity orbit, the loss of warmth at the far end of the ellipse is exactly balanced by the gain in warmth at the closest approach to the sun, compared with the heat that would have been received if the Earth had a perfectly circular orbit. But, as Croll saw, the changing eccentricity does alter the balance of heat received by the two hemispheres of the Earth in different seasons.

When the orbit is nearly circular, there is no eccentricity effect on the seasons. But when the orbit is more eccentric, in one hemisphere winters are particularly cold, because the Earth is farthest from the sun, and summers are particularly warm, because the Earth is closest to the sun, while in the other hemisphere there is less difference between the seasons. Croll guessed that in order to make an ice age the world would need a succession of very cold winters in the northern hemisphere, so that snow and ice would build up over the land around the Arctic Ocean. He realized that, on its own, the eccentricity effect was not strong enough to do the job, So he suggested that as the white snowfields built up they would reflect away some of the summer heat from the sun, keeping the northern hemisphere cool and producing a chain of positive feedback that encouraged the spread of ice. Croll was one of the first scientists to discuss the idea of feedback in any context, and this is a key part of the modern understanding of ice age rhythms. He also noted that the way the Earth wobbles (like a wobbling spinning top) as it orbits the sun also plays a role in the story, as this affects the pattern of the seasons. This is the origin of the other two cycles that are part of the ice age rhythm. But as it happens, he got the basic concept completely backwards.

According to Croll's calculations, there should have been an ice age that peaked about one hundred thousand years ago, at a time of high orbital eccentricity, followed by a long interglacial that occurred when the eccentricity was lower. But as more geological evidence was gathered, and dating techniques improved, by the end of the nineteenth century it was clear that the latest ice age had ended by about ten thousand years ago, and that the previous warm interval, an interglacial akin to present-day conditions, had occurred roughly one hundred thousand years ago, exactly when Croll had thought the ice age should be at its peak.

The astronomical model of ice ages fell out of favour as a result of these discoveries – because nobody thought at the time to ask why Croll's model was *exactly* the opposite of what is actually happening to the Earth. What we now understand is that in order for ice sheets to spread in the northern hemisphere we need a succession of cold *summers*. Snow always falls in winter, but what matters is how much, or how little, of it stays through the summer season. If less snow melts in summer than falls in winter, the feedback effect that Croll described can get to work, with the result that the ice sheets can build up very rapidly. In fact, with the present distribution of the continents, the natural state of the Earth is in a full ice age, and it is only tugged out of the ice age and into an interglacial when all of the astronomical influences conspire to produce hot summers in the northern hemisphere, melting the ice back part of the way to the pole. The southern hemisphere, of course, is in a permanent ice age because Antarctica sits squarely over the South Pole.

The person who re-established the astronomical model of ice ages as a good explanation of the way the real world behaves was a Serbian, Milutin Milankovitch, who did such a good job that the idea is now often referred to as the 'Milankovitch Model' of ice ages.

Milankovitch thought big. He didn't just set out to study the changing climate of the Earth, but had the ambition of calculating the temperature changes over the past million years at different latitudes on each of the three planets Venus, Earth, and Mars. And he set out on this self-imposed task in 1911, at the age of thirty-two, armed only with paper, pen, and his own brain. At the time, he was professor of applied mathematics at the University of Belgrade; the task occupied him for the next thirty years.

His first breakthrough came in bizarre circumstances. In 1914, when World War I broke out, Milankovitch was on the wrong side of the frontier, on a visit to his hometown of Dalj, and was imprisoned by the Austro-Hungarian authorities. After a few months in jail, because of his academic status he was allowed to spend the rest of his internment living in Budapest and working in the library of the Hungarian Academy of Sciences, only required to report to the police once a week. For the next two years, he was able to work uninterrupted on his calculations, coming up with a complete set of equations (a mathematical model) describing the climate on Earth today. When the war ended, he spent the next two years adapting the model to describe the climates of Venus and Mars, then published his results in a book.

Few people took much notice at the time, but over the years and decades that followed, Milankovitch refined his calculations, extending them to more latitudes on Earth and applying them, at the suggestion of the meteorologist Wladimir Köppen, to calculate when northern hemisphere summers were particularly cold. His complete model was published in 1941, in another book. By then, the problem was that the mathematical calculations were far more accurate than the geological data! Milankovitch could say exactly when, at any time in the past million years, the Earth should have been in an ice age or in an interglacial. But in the

1940s and 1950s the geologists couldn't make measurements of past climate accurate enough to test his numbers. That only began to happen in the 1960s, when scientists began to probe Earth history using cores of sediments drilled from the deep ocean, work that came to full fruition in the 1970s.

The core of climatic change

Past climates can be studied using cores from the deep sea because there are two kinds of oxygen in the air we breathe. They are called isotopes, and are known as oxygen-16 and oxygen-18, where the numbers refer to the weight (strictly speaking, the mass) of each kind of oxygen atom. An atom of oxygen-18 is two units heavier than an atom of oxygen-16, but both have the same chemical properties. Both kinds of oxygen atom can combine with hydrogen atoms to make molecules of water, H_2O. But water molecules that contain oxygen-18 are two units heavier than water molecules that contain oxygen-16, and as a result they do not evaporate from the sea as readily as water molecules that contain oxygen-16. The ratio of oxygen-18 to oxygen-16 in water molecules in the air, and the corresponding ratio for the water molecules left behind in the sea, depend on the average temperature of the globe. By measuring either of these ratios, it is possible to get a very accurate measure of the temperature.

Both kinds of measurement have been used in practice. When water molecules from the air fall as snow, they carry with them the isotopic 'signature' of the global mean temperature at the time, and this is preserved in layers of snow that fall each year and build up into ice sheets. When cores are drilled from the ice, the layers can be dated by counting them down from the top, just as tree rings can be dated by counting them in from the bark, and the average temperature on Earth at different times in the past can be calculated from the isotope ratios. But this was not the technique that led to the rehabilitation of the Milankovitch Model.

That breakthrough came from studies of the ratio of the isotopes left behind when huge quantities of snow fell and built up in this way to make the glaciers of an ice age. This means there is relatively less oxygen-18 in the ice, and relatively more of it in the sea, during an ice age. So the tiny marine creatures called plankton collect a higher proportion of oxygen-18 in their shells when the world is colder, because they take up oxygen from the water while they are alive. When they die, their shells fall to the bottom of the sea and accumulate there in sediments. It is much harder to obtain a core of sediment from the deep sea than it is to obtain a core from an ice sheet on land, and it is much harder to date the sediments accurately, because they do not form distinct annual layers like the layers of ice on land. But from the 1950s onwards geophysicists developed improved methods of radioactive dating that could be applied to these sediments, and better-equipped drilling ships became operational.

By the late 1960s, various pieces of evidence, from both land and sea, began to hint that the Milankovitch Model might be correct; but there was no single, definitive proof. So in 1971 the US National Science Foundation agreed to fund a project called CLIMAP (from the words 'Climatic Mapping') to investigate how the climate of the North Pacific and the North Atlantic has varied over the past seven hundred thousand years, the interval since the latest reversal of the Earth's magnetic field. The project soon hit the jackpot with a core prosaically labelled V28-238, but which soon became known to geophysicists and climatologists as the 'Rosetta Stone'.

The Rosetta Stone core came from the western equatorial Pacific, and provided a complete record going back beyond the seven-hundred-thousand-year target date. The isotope variations in the core clearly showed the one-hundred-thousand-year cycle of ice ages and interglacials that had been discussed since the second half of the nineteenth century, plus some less dramatic fluctuations that needed more careful analysis. The overall effect is to produce a complicated pattern that looks like the jumble of sound waves produced when a chord is struck on a guitar. But just as that pattern of sound waves is made up of several single notes adding together, so the pattern of temperature changes is made up of several climate cycles adding together.

Separating out the single 'notes' of the climate cycles is actually a fairly simple mathematical problem, and the Rosetta Stone core did indeed contain evidence of the roughly forty-thousand-year-long and twenty-thousand-year-long cycles also predicted by the Milankovitch Model. Another core, from the bed of the Indian Ocean, provided an even more detailed record. Although it only extended back in time for just under half a million years, it showed clear evidence of all the Milankovitch cycles. In 1976, all of this evidence was gathered together and published in a scientific paper in the journal *Science*, under the title 'Variations in the Earth's Orbit: Pacemaker of the Ice Ages'. This was the precise moment when the Milankovitch Model came in from the cold. Since then, a wealth of other evidence, including data from ice cores, has confirmed its accuracy. And the proof that this rhythm of repeating ice ages and interglacials has persisted for millions of years has shed new light on human origins.

10
Life and Earth

The earliest evidence for life on Earth comes from fossilized organic remains found in rocks from Greenland that are 3800 million years old. But there is no suggestion that life started at that time, and it is likely that there was already life on our home planet more than 4000 million years ago.

The first large structures associated with living things were mounds produced in shallow water by the growth of mats of algae, simple, single-celled organisms. Algae mats trap mud particles, another layer of algae forms on top of the mud, another layer of mud forms on top of the algae, and so on. The resulting mounds, which can be a metre or so high, are called stromatolites, and fossil stromatolites show that this kind of life was flourishing in warm, shallow waters by around 3500 million years ago.

That mention of warm, shallow water is a key to the story of life on Earth, since life as we know it requires the existence of liquid water. Venus, the next planet in towards the sun from Earth, is too hot for liquid water to exist; Mars, the next planet out from the sun from the Earth, is too cold to have oceans. Both seem to be lifeless. But on Earth the temperature, like that of Baby Bear's porridge in the Goldilocks tale, is just right, and there is an abundance of both liquid water and life. (Astronomers can calculate where the 'life zone' in which liquid water can exist on a planet lies for stars of different masses, which gives them a clue where to look for 'other Earths', but the story of that search is beyond the scope of our present book.)

We know from fossils that more complex cells with a central nucleus, called eukaryotes, had evolved on Earth by about

2500 million years ago, and it was some of these organisms that 'invented' photosynthesis, gradually releasing oxygen into the atmosphere. Even at a much lower concentration than in the present atmosphere, oxygen formed an ozone layer high above the planet's surface, and this ozone filtered out solar ultraviolet radiation that is damaging to DNA, which allowed life to flourish more widely. These photosynthesizers in essence inherited the Earth. Yet they remained microscopic, single-celled organisms until about 1200 million years ago.

What followed has been described as the 'big bang' of evolution – the emergence of multicellular organisms and the development of sexual reproduction. Together, these changes led to a diversification of life in the sea, larger organisms, the division of life into plants and animals, and, by a bit less than 600 million years ago, the appearance of creatures with shells. Because shells make much better fossils than soft bodies do, it's from that time that the story of life on our Earth, and emergence of these life forms from the oceans onto the land, can be understood in more detail. This happened at the end of an interval of geological time – roughly eighty-five per cent of Earth history – known as the Precambrian.

From the Precambrian to prehumans

Among the last of the Precambrian life forms there was a group of organisms, known as ediacarans, that lived in the sea around 550 million years ago; they superficially resembled jellyfish. Some were flat discs, others were up to a couple of metres long and had more complicated body shapes, with a distinct 'head' and 'tail'. Although they were soft-bodied, the imprints of the remains of about a hundred species of ediacarans have been found, as fossilized moulds of the sediments which surrounded them during their lives.

The interval of geological time that followed the Precambrian is called the Palaeozoic era, and is subdivided into periods, the first of which is called the Cambrian (which is how the Precambrian got its name). Gondwana lay in the southern hemisphere early in the Palaeozoic, gradually moving north and linking up with other continental crust to form Pangaea. At first, these geographical conditions favoured life in the shallow seas around the landmasses, where coral reefs flourished, and it was during the Cambrian that most of the main groups of marine invertebrates evolved, including the trilobites, which are now extinct but once proliferated and have left many fossils behind. From our human point of view, an even more significant development was the evolution of the first fish-like creatures with the beginnings of a backbone, a rod called the notochord. All vertebrates, including ourselves, evolved from creatures like these.

Another significant step towards our own existence came when life began to move out of the sea and on to the land. Fossil footprints and plant spores show that this happened by the Ordovician, the geological period that followed the Cambrian and lasted from about 495 million years ago until about 443 million years ago. Both plants and creatures like millipedes and insects, collectively known as arthropods, spread across the land in the Silurian, from about 443 million years ago until about 417 million years ago. But it was only in the Devonian period, between 417 million years ago and 354 million years ago, that our direct ancestors, a kind of proto-amphibian, waded ashore. They lived in lush forests covering extensive regions of low-lying wetlands. Although these creatures had to return to the water to reproduce, during the following period, the Carboniferous, both amphibians proper and the first reptiles evolved. The key evolutionary development, of course, was the ability of reptiles to lay eggs protected by an outer covering, so that they did not have to return to the water to breed. Early reptiles were two or three metres, or six to ten feet, long, and resembled large lizards.

The Carboniferous gets its name because it was during this period, from about 354 million years ago until about 290 million years ago, that great forests flourished, taking carbon dioxide out of the air and storing the carbon from it in the trunks of trees which then became fossilized and turned into layers of coal – the fuel that powered the industrial revolution (which then put much of this carbon dioxide back into the air, producing problems with our climate today). So much was going on in the Carboniferous (including the evolution of seed-bearing plants) that it is sometimes subdivided into the Mississippian, from 354 million years ago to 323 million years ago, and the Pennsylvanian, from 323 million years ago to 290 million years ago, but this subtlety is not important to the broad sweep of our story.

During the following period, the Permian, which lasted from about 290 million years ago to about 248 million years ago, reptiles diversified prolifically, thanks to that key invention of the waterproof egg. Pangaea became the home to the ancestors of dinosaurs, turtles, and mammals, as well as many other short-lived variations on the reptilian theme.

But at the end of the Permian, some 248 million years ago, everything changed dramatically, which is one reason why some of those variations were so short-lived. Then, something happened on Earth to bring about the extinction of ninety per cent of all species, encompassing sixty per cent of all genera, the level of taxonomic classification one step up from species. This is the largest mass extinction known from the fossil record, an event so spectacular that it literally marks the end of an era, the Palaeozoic.

Nobody knows exactly what happened to cause the disaster. At the time, Pangaea stretched from pole to pole, encouraging ice sheets to form and lowering sea level, but this alone could not have been responsible. The best indication of a 'smoking gun' comes from evidence found in the huge lava fields in Siberia, known as the Siberian Traps, which were produced in a vast outburst of volcanic activity at the end of the Permian period. This activity created a layer of lava three kilometres (about two miles) thick over an area of about 2.5 million square kilometres (one million square miles), all erupted in less than a million years. Such an eruption must have spread enormous amounts of dust, carbon dioxide, and sulphur dioxide into the atmosphere, where the sulphur dioxide would react to make sulphuric acid droplets. The resulting acid rain alone could account for the death of things like corals, trilobites, and other forms of life in the sea; the dust and sulphuric acid droplets high in the atmosphere would block the heat from the sun and cool the globe into a short-lived ice age; while in the longer term the injection of huge quantities of carbon dioxide into the atmosphere caused an increased greenhouse effect, raising temperatures well above present levels before they settled back down again.

But the survivors of such extinctions have an opportunity to spread out and diversify as the world recovers from the shock. That is exactly what happened during the Mesozoic era, which spanned some 183 million years and was made up of three periods: the Triassic, from 248 million years ago to 206 million years ago; the Jurassic, from 206 million years ago to 142 million years ago; and the Cretaceous, from 142 million years ago to sixty-five million years ago. This was the age of the reptiles, and from the end of the Triassic to the end of the Cretaceous, roughly a span of 140 million years, it was the age of the dinosaurs. To put that in perspective, the direct evolutionary line leading to ourselves split off from the other apes no more than five million years ago; we have been around for less than four per cent of the time that the dinosaurs dominated the Earth.

Although life also flourished and diversified in the sea during the Mesozoic, with creatures such as turtles, whose descendants are around today, evolving alongside creatures such as ichthyosaurs, which are now extinct, our focus now turns to the land, where our own ancestors lived. In the early Triassic, our direct ancestors were warm-blooded, semi-reptilian creatures known as therapsids. They were in direct competition for ecological niches and resources with the more reptilian diapsids. This was a genuine clash between two variations on an evolutionary theme. It is a sobering thought that in this clash, the therapsids, our ancestors, came second. The diapsids evolved into dinosaurs, which would dominate the Earth, while many therapsid species went extinct. The therapsids were not all wiped out, but the ones that survived did so by evolving into small, mouse-like creatures, too insignificant to bother the dinosaurs. They probably led a nocturnal lifestyle and lived off insects and plants.

So it was that dinosaurs occupied most of the land-animal habitats of the Mesozoic, and also claimed the air and the sea. There were dinosaur equivalents of sheep, lions, elephants, sharks, and just about any kind of large mammal that you can think of. In fact, one branch of the dinosaur line evolved into birds, and in this sense it could be said that the dinosaurs never went extinct. But the age of the dinosaurs ended some sixty-five million years ago, when another great catastrophe struck the Earth, killing fifty per cent of all species and bringing an end to the Cretaceous era.

At the end of the Cretaceous, life on Earth was hit by at least a double, and possibly a triple, whammy. By the time of the extinction, called the Terminal Cretaceous Event, dinosaurs and other forms of land-based life had been in decline for millions of years. Because of the tectonic changes we have described, shallow seas, such as the remains of the Tethys Ocean, were shrinking and continental interiors were drying out, producing a harsher climate of hot summers and cold winters. Then, sixty-five million years ago, one or two dramatic events occurred that combined with these slower changes to make a bad situation worse and wipe out fifty per cent of species.

One of the candidates for this 'last straw' effect is the impact of a large rock from space, a piece of cosmic rubble measuring about ten kilometres, or six miles, across. We know for sure that there was such an impact at the end of the Cretaceous. The impact site itself has been identified, beneath what is now the Yucatan Peninsula in Mexico, and there is a thin layer of debris from the impact found in sediments of the right age around the world. Any such impact would cause a short-lived global cooling because of the dust and other debris spread high into the atmosphere. But this particular impact also happened to strike a region of rocks rich in sulphur and carbonates.

Heat from the impact caused the production of huge quantities of sulphur dioxide, which made sulphuric acid rain. Limestone rocks in the same region released carbon dioxide to warm the world through an enhanced greenhouse effect. The overall result would be a downpour of acid rain, a mini-ice age, and then a longer interval of global warming. There is geological evidence for all of these effects around sixty-five million years ago.

Does the pattern look familiar? It should. Overall, the effect of this particular impact was like a smaller-scale version of a volcanic outpouring like the one that formed the Siberian Traps. Curiously, around sixty-five million years ago, on the other side of the world from the impact, there was another volcanic outburst of this kind, producing what are known as the Deccan Traps of central India. Was it just bad luck that a population already weakened by gradual climate change was hit at the same time by an impact from space and by a huge upsurge of volcanism? Although it can only be speculation, some researchers suggest that this is no coincidence at all, but that the shock waves from the impact, rippling around the world, may actually have been the trigger for the volcanic activity. But whatever the details of the Terminal Cretaceous Event, it did bring an end to the age of the dinosaurs, opening the way for the age of the mammals.

Our ancestors survived the death of the dinosaurs because they were small. This meant that they could hide in tunnels or burrows from the worst effects of the Terminal Cretaceous Event, and that they did not need huge quantities of food in order to get

by during harsh times. Of course, mammals weren't the only
beneficiaries of the opportunities that opened – once conditions
for life on Earth improved – because of the disappearance of the
dinosaurs and many other Cretaceous species. The Caenozoic
era, from sixty-five million years ago up to the present day, was
characterized, until very recently, by a diversification not only of
mammals but of birds, bony fish, and flowering plants, as well as
the insects that pollinate them. While life was evolving, the face
of the Earth was also changing, as the continents slid into their
familiar positions; the great mountain regions of the Andes, the
Rockies, the European Alps, and the Himalayas are all products
of the Caenozoic, too.

Human beings, *Homo sapiens*, are members of a biological
order called the primates. This group of animals is characterized
by having opposable thumbs (and in many cases opposable big
toes), which allow for great dexterity, and forward-facing eyes,
which provide three-dimensional vision. These capacities go
hand in hand with the development of large brains. Monkeys and
apes are closely related members of the primate order, but belong
in different biological superfamilies. Modern Old World monkeys and apes (including ourselves) are descended from a common
line that split to create these two groupings about thirty million
years ago, perhaps a little less. The ape superfamily is known as
the *Hominoidae* or hominoids, and is split into two families, the
lesser apes (the gibbons) and the great apes.

The great ape family, known as *Hominidae* or hominids, contains only ourselves and other apes – by all the rules of biological
classification, humans are apes. For some twenty million years,
during the Miocene epoch, conditions were reasonably calm, and
the apes, which had evolved in Africa, had an opportunity to
diversify and spread. Many species of ape that are now extinct
evolved, some leaving Africa to populate large areas of Eurasia as
the two continents were joined together. The early Miocene in
particular was a time when mammals in general came into their

own, as the steady drift of the continents brought a shift to a drier climate that encouraged grassland to spread at the expense of forest, and animals to spread around the globe in great waves of migration. If any geological interval of time resembles the description of the mythological Garden of Eden, this was it.

But the good times ended a little over five million years ago, as ice began to form over both poles. It was just after this that our own line, leading to the species *Homo sapiens*, split from the other apes – and the timing is no coincidence.

Winter in Eden

Our closest living relatives are three species, two of chimpanzee plus the gorilla. Evidence from DNA shows that we share a common ancestor with these species that was alive between four million and five million years ago. The technique is sufficiently robust to tell us that the gorilla line split off first, and the human–chimpanzee split occurred a little later. The chimpanzee line itself then split, around two million years ago, into the smaller, forest-dwelling pygmy chimpanzee, *Pan paniscus*, also called the bonobo, and *Pan troglodytes*, the larger inhabitant of the wooded savannah that we know simply as the common chimpanzee. All of this occurred against a background of climate that wasn't just changing, but pulsing, thanks to the unusual arrangement of the continents, with the Milankovitch rhythm.

What did this repeating rhythm of ice ages and interglacials mean for these African forest dwellers? To them, the Milankovitch cycles were not so much a rhythm of warmer and colder temperatures, but a regular succession of wet and dry conditions. During a full ice age, water is locked up in the ice and the sea levels fall, while the oceans cool and there is less evaporation. Moisture-bearing winds contain less moisture, and they have farther to travel to reach the interiors of continents. During an interglacial, the rains return, to some extent. So the African forests shrink during an ice age, but expand during an interglacial.

Changing climate is one of the factors that affect the evolution of species. As Charles Darwin and Alfred Russel Wallace realized back in the nineteenth century, evolution works by a process of natural selection. In each generation, there are individuals who are slightly different from one another. We now know that this is because of differences in their genes, which are made of DNA. To take a simple example, one gazelle in a herd may be able to run faster than another gazelle in the same herd. If one individual is better suited to its environment (better 'fitted' in the way a piece of jigsaw puzzle fits in to the overall picture), it is more likely to survive and have offspring – in our example, the gazelle that can run faster is more likely to escape from predators. Crucially, the genetic material is passed on to the next generation, and the individuals that are best fitted to their environment have more offspring, so their genes become more common in the next generation. This is what is meant by the term 'survival of the fittest'. It doesn't necessarily mean 'fit' in the sense of athletic fitness, although that certainly can help, but could apply to, for instance, intelligence, or the possession of good three-dimensional vision. And the only measure of evolutionary success, as a species, is whether some individuals survive long enough to have offspring; if they don't the species goes extinct. (It's worth pointing out that evolution is a fact, like the fact that apples fall down off trees, and natural selection is a theory that explains the fact, just as the theory of gravity explains why apples fall down.)

As long as the environment stays the same, evolution does a job of fine-tuning, which makes species better and better adapted to their ecological niches. But when the environment changes (or when individuals move into new environments, like the mammals spreading out across the globe) something that used to be unimportant for survival may become just the thing that individuals need. Then, evolutionary pressures can lead to the emergence of new species.

There are two ways in which evolutionary pressures of this kind will affect forest-dwelling apes when the forests shrink. One solution to the problem is to stick tight to the remaining trees, and become even more adept at living in the forest, better fitted to the old environment – in common language, to become more 'ape like'. The other solution is to adapt to life on the edge of the forest, or even outside the forest on the savannah, developing a more flexible lifestyle and becoming non-specialists, good at getting by on whatever is available – in common language, to become more 'human'.

But this takes time. If the forests shrink as a result of a permanent change in climate, the best-adapted apes will cling on in the forests, and their less successful relations on the edges of the forest will die. The first of the gang to die will be the ones that are least flexible in their lifestyle and least good at finding new sources of food, but they will soon be followed by their slightly more adaptable cousins. In each generation, the most adaptable individuals will survive and pass on their genetic material, including whatever it is that makes them adaptable, but in the long run it will do them no good, as the population declines in each generation.

If, however, after a hundred thousand years of this, the rains return and living becomes easier, the survivors, the inheritors of those genes for flexibility, will thrive and their population will boom. When the forests shrink again and life becomes harder, there will be a large enough population of adaptable creatures for the species to survive the same winnowing process again. Over many ice age–interglacial cycles, this repeated winnowing will select strongly for the kind of versatility and adaptability that makes us human. So it is a combination of the Milankovitch astronomical cycles and the present geography of the globe produced by continental drift that has made us human.

There is one other thing to take on board about this process. The *successful* woodland apes stayed in the woods all this time. The *unsuccessful* woodland apes were forced out of the forest and had to become more adaptable to survive. We are descended from the unsuccessful apes.

Actually, that kind of reasoning applies on a much larger scale. Way back in geological history, the successful fish who were good at being fish stayed in the oceans; unsuccessful fish were forced to find a new way of life, on land. Similarly, when successful amphibians filled all of the ecological niches available to them, unsuccessful amphibians had to 'learn' to become reptiles – and so on. The most successful forms of life on Earth are the single-celled algae that have been around for billions of years, beautifully adapted to their environment. We are descended from a long line of creatures forced to find new ways of life by the success of other creatures. In a sense, we are failed algae.

But that is ancient history. On the scale of the present ice epoch, looking back over the past five million years, we can see just how the human line emerged from the primate family of hominids. The modern genus of humans, known as *Homo*, traditionally contains only one living species, ourselves, and those of our fossil ancestors that came after the human–great ape split. This classification is an example of extreme human chauvinism, and recently there has been a move to include the other African apes and orangutans in the same category. In fact, by all the usual rules we ought to be placed in the same genus as the chimpanzees, perhaps as *Pan sapiens*, but that certainly isn't likely to happen in the near future.

The earliest hominid that is given the genus name *Homo* evolved in East Africa by about three million years ago, thanks to the evolutionary pressures we have described, in the region of the Great Rift Valley. Dubbed *Homo habilis*, this variation on the ape theme stood about 1.2 metres – four feet – tall, had a slight build, and a brain averaging 675 cubic centimetres in volume, about half that of modern *Homo sapiens*. By about 1.5 million years ago, the continuing ice age–interglacial winnowing had produced *Homo erectus*, the direct descendant of *H. habilis*, who was taller, at about 1.6 metres, or five feet three inches, and had a brain with a volume of about 925 cubic centimetres. It was *H. erectus* who spread the hominid line out of Africa and into Asia. Our own species, *Homo sapiens*, had evolved from *H. erectus* by about half a million years ago.

Soon after the emergence of *Homo sapiens*, the line split into two – *Homo sapiens sapiens* and *Homo sapiens neanderthalensis*, or Neanderthal man. The two sub-species lived on the same planet, though not necessarily alongside each other, from at least one hundred thousand years ago (the start of the most recent ice age) until about forty thousand years ago. The Neanderthals seem to have been better adapted to the colder conditions in the north, and may have been unable to survive the combination of a warming world and the competition from *H. sapiens sapiens* when the latest interglacial got under way – or, some scientists suggest, they may have interbred with *H. sapiens sapiens* and disappeared in part that way.

Geologically speaking, the present interglacial is not significantly different from the ten or twenty that preceded it. But it is the one during which humankind has spread around the globe, making such an impact that, if there are any palaeontologists around in the far future to study the fossil record, they will surely make it out to be the end of an era, marked by another great wave of extinctions.

The sixth extinction

Although there have been many occasions during the long history of the Earth when extinctions large enough to leave a mark in the fossil record have occurred, five stand out as really dramatic events in the history of life. Known as the 'Big Five', these events are the most striking boundary markers in the geological record. The first occurred about 440 million years ago, and marks the boundary between the Ordovician and Silurian periods of geological time. The second occurred some 355 million years a go, marking the end of the Devonian. Another, the biggest of them all, occurred about 250 million years ago, at the end of both the Permian period and the Palaeozoic era, and the fourth of the Big Five extinctions was the one that occurred a little more than 200 million years ago, at the end of the Triassic. The fifth, as we have just described, occurred sixty-five million years ago, bringing an end both to the Cretaceous period and the age of the dinosaurs.

These events stand out because they are unusual, with a large number of species disappearing in a very short span of geological time. But extinction is a way of life on Earth, and the geological record shows us that even in the quiet intervals between extinctions, about one species dies out every four years. Given the huge number of species, this is equivalent to saying that the average lifetime of a species is a few million years, before it either dies out completely or evolves into something else, in the way some dinosaurs evolved into birds, and *Homo erectus* evolved into *Homo sapiens*.

But that steady, background rate of extinction is not what is happening now. Experts, including the American biologist E.O. Wilson, have estimated that species are disappearing today at a rate of about fifty thousand *per year*. That is two hundred thousand times faster than the 'normal' quiet rate of extinctions between catastrophic events. It means that we are living through another great extinction, turning the Big Five into a Big Six. What makes this sixth extinction unique is the rate at which it is happening – species are disappearing much faster than during any of the previous big extinctions, compressing the kind of disaster that is usually measured in millions of years into centuries. It is estimated that fifty per cent of all the species of life on

Earth that were around before this process started will have died out by the end of the twenty-first century, making this a catastrophe to rank with the disaster that killed off, among other things, the dinosaurs.

We don't have to look far for the cause. This time, it isn't the eruption of a huge volcanic outburst lasting for hundreds of thousands of years, or the sudden impact of an asteroid, that is to blame, but our very selves. The sixth extinction under way is almost entirely a result of human activities, including deforestation, the conversion of natural habitat such as prairie into farmland, and, increasingly, the effects of global warming, killing off temperature-sensitive species such as coral.

Humankind, in other words, has become a geological force. Another way of looking at this is suggested by James Lovelock, who proposed that everything on Earth, including both living and non-living systems, behaves like the components of a single organism, which he calls Gaia. (Scientists who are uncomfortable with the implied anthropomorphism prefer the name Earth System Science, which means the same thing.) This system resembles a kind of biological cell, and Lovelock, who is a qualified physician, describes humankind as like a virus infecting the planet-cell and disrupting its natural workings. He has no doubt that Gaia will survive, just as she has survived the previous five great extinctions. But he suggests that the best way for Gaia to survive might be to rid herself of us. Lovelock's prognosis is bracing, but it seems to be exactly the sort of harsh medicine we need to swallow, or ignore at our own peril.

If planet Earth is to continue to be a comfortable home for humankind, we are going to have to learn to live in harmony with the natural systems that have allowed life to flourish here for thousands of millions of years. And we are going to have to learn that lesson soon.

Appendix 1
The Earth in numbers

Age of the Earth (approx.): 4.5 billion years

Distance to the sun (average): 149,669,180 kilometres/93,000,116 miles

Distance to the moon (average): 384,403 kilometres/238,856 miles

Orbital period around the sun: 365.256 days

Orbital speed (average): 29.78 kilometres per second/18.50 miles per second

Rotation: 23 hours, 56 minutes, 4.09 seconds

Circumference at the equator: 40,075 kilometres/24,902 miles

Diameter at the equator: 12,756 kilometres/7926 miles

Diameter at the poles: 12,714 kilometres/7900 miles

Radius (average): 6371 kilometres/3959 miles

Diameter of the inner core (approx.): 2432 kilometres/1511 miles

Surface area: 510,072,000 square kilometres/196,939,900 square miles

Density of the planet (approx.): 5520 kilograms per cubic metre/0.1994 pounds per cubic inch

Density of the core (approx.): 12,000 kilograms per cubic metre/0.4335 pounds per cubic inch

Temperature at the surface (average): 15 degrees Celsius/59 degrees Fahrenheit

Temperature at boundary between inner and outer core (approx.): 5505 degrees Celsius/9941 degrees Fahrenheit

Altitude of Kármán line, the boundary between atmosphere and outer space (approx.): 100 kilometres/60 miles
Altitude of Mount Everest: 8848 metres/29,029 feet
Altitude of Denver, USA: 1731 metres/5679 feet
Altitude of London, UK: 24 metres/79 feet

Depth of Dead Sea: 418 metres/1371 feet below sea level
Depth of the Mariana Trench, Pacific Ocean, at Challenger Deep: 10,911 metres/35,797 feet below sea level
Depth of the boundary between the inner and outer core (approx.): 5100 kilometres/3100 miles

Human population (2010, approx.): 6.82 billion

Appendix 2

The timescales of planet Earth

By studying rock strata, geologists have divided Earth history into units of time. Eons are the longest unit in this timescale, lasting for about half a billion years, with six eons in Earth history. Eons are divided into eras of several hundred million years, and these eras are subdivided into periods, which are themselves divided into epochs. So the history of our planet can be told using an approximate timeline.

Precambrian super-eon		
Approx. 4.5 billion to 550 million years ago		
Hadean eon	**Archaean eon**	**Proterozoic eon**
Earlier than 3.8 billion years ago	3.8 billion to 2.5 billion years ago	2.5 billion to about 550 million years ago
Earth bombarded by 10 billion billion tonnes of material from space	*High volcanic activity and several small continents moving quickly through tectonics; emergence of algae mats*	*Larger continents form, at times in a single supercontinent; massive flood basalt eruptions; emergence of eukaryotes through to ediacarans; Pangaea supercontinent forms near end of eon*

Phanerozoic eon				
Palaeozoic era 550 million to 248 million years ago				
Cambrian period	**Ordovician period**	**Silurian period**	**Devonian period**	**Carboniferous period**
550 million to 495 million years ago	495 million to 443 million years ago	443 million to 417 million years ago	417 million to 354 million years ago	354 million to 290 million years ago
Gondwana continent in southern hemisphere; coral and other marine life – animals with shells and bones – flourishes	*Life moves on to the land; ends with mass extinction event*	*Plants and insects spread across landmasses*	*Pangaea forms over the South Pole, then breaks up; proto-amphibians emerge; ends with mass extinction event*	*Amphibians and reptiles emerge, great forests flourish*

Phanerozoic eon			
Palaeozoic era 550 million to 248 million years ago	**Mesozoic era** 248 million to 65 million years ago *Age of the dinosaurs*		
		Triassic period 248 million to 206 million years ago *Pangaea single supercontinent forms again; diapsids (ancestors of dinosaurs) and therapsids (ancestors of mammals) emerge*	**Jurassic period** 206 million to 142 million years ago *Break up of Pangaea into small continents; dinosaurs dominate Earth*
Permian period 290 million to 248 million years ago *Reptiles proliferate; ends with mass extinction event – the largest in Earth history*			**Cretaceous period** 142 million to 65 million years ago *Continents begin to take modern shape and interiors dry; dinosaurs dominate Earth; ends with mass extinction event – the end of the dinosaurs*

Phanerozoic eon			
Caenozoic era 65 million years ago to present day *Continents take current positions; diversification of mammals, birds, fish, insects, and flowering plants*			
Paleogene period 65 million to 23 million years ago			
Paleocene epoch 65 million to 56 million years ago	**Eocene epoch** 56 million to 34 million years ago *Supercontinent Laurasia breaks up and the remnants of the Tethys Ocean disappear; modern mammals emerge*	**Oligocene epoch** 34 million to 23 million years ago *Expansion of grasslands; Amazon river system created*	

Phanerozoic eon			
Caenozoic era 65 million years ago to present day *Continents take current positions; diversification of mammals, birds, fish, insects, and flowering plants*			
Neogene period 23 million to 2.5 million years ago		Quaternary period 2.5 million years ago to present day	
Miocene epoch 23 million to 5 million years ago *Mammals flourish and apes spread*	**Pliocene epoch** 5 million to 2.5 million years ago *First hominins appear*	**Pleistocene epoch** 2.5 million to 11,700 years ago *Glaciation cycles; Homo sapiens emerge around 500,000 years ago*	**Holocene epoch** 11,700 years ago to present day *Current interglacial begins; growth of human civilization*

Further reading

Cattermole, Peter (2000) *Building Planet Earth*. Cambridge: Cambridge University Press.

Darwin, Charles (1842) *The Structure and Distribution of Coral Reefs*. London: Smith, Elder & Co. Available online at http://www. darwin-literature.com/Coral_Reefs/.

Drury, Stephen (1999) *Stepping Stones*. Oxford: Oxford University Press.

Fortey, Richard (2004) *The Earth*. London: HarperCollins.

Gribbin, John (2002) *Science: A History*. London: Penguin.

—— (2000) *Stardust*. London: Penguin.

Gribbin, John, and Mary Gribbin (2008) *From Here to Infinity*. London: National Maritime Museum, Royal Observatory Greenwich.

—— (2001) *Ice Age*. London: Allen Lane.

Hooke, Robert (1665) *Micrographia*. London: Royal Society.

Imbrie, John, and Katherine Palmer Imbrie (1979) *Ice Ages: Solving the Mystery*. London: Macmillan.

Leakey, Richard, and Roger Lewin (1996) *The Sixth Extinction*. London: Weidenfeld & Nicolson.

Levy, Matthys, and Mario Salvadori (1995) *Why the Earth Quakes*. New York: Norton.

Lovelock, James (1991) *The Practical Science of Planetary Medicine*. London: Gaia Books.

Luhr, James (ed.) (2004) *Earth*, 2nd edn. London: Dorling Kindersley.

Miller, Russell (1983) *Continents in Collision*. Amsterdam: Time-Life Books.

Owen, H.G. (1983) *Atlas of Continental Displacement, 200 Million Years to the Present*. Cambridge: Cambridge University Press.

Ritchie, David (1981) *The Ring of Fire*. New York: Atheneum.

Stewart, Iain (2005) *Journeys from the Centre of the Earth*. London: Century.

Stewart, Iain, and John Lynch (2007) *Earth*. London: BBC.

Wilson, E.O. (2002) *The Future of Life*. New York: Alfred A. Knopf.

Winchester, Simon (2003) *Krakatoa*. London: Viking.

Wood, Robert Muir (1985) *The Dark Side of the Earth*. London: Allen & Unwin.

Acknowledgements

We are grateful to the University of Sussex, which provided us with a base to work from, and the Alfred C. Munger Foundation, which contributed to our travel and other expenses as we researched and wrote the book.

Index

A Beginner's Guide to History of Science

9781851686810
£9.99/ $14.95

Sean Johnston weaves together intellectual history, philosophy, and social studies to offer a unique appraisal of the nature of this evolving discipline. This book demonstrates that science is a continually evolving activity that both influences and is influenced by its cultural context.

"Lucidly and engagingly written ... Johnston has managed to cover an impressive range of material, making it readily accessible to newcomers." **Patricia Fara** – author of *Science: A Four Thousand Year History*

"Clearly written without being patronising, this is a first-rate introduction to the history of science! " **Dr Peter Morris** – Head of Research at the Science Museum, London

SEAN F. JOHNSTON is Reader in the History of Science and Technology at the University of Glasgow. He is also a Fellow of the Higher Education Academy with a prior career as a physicist and systems engineer.

Browse further titles at
www.oneworld-publications.com

A Beginner's Guide to Quantum Physics

9781851683697
£9.99/ $14.95

From quarks to computing, this fascinating introduction covers every element of the quantum world in clear and accessible language. Drawing on a wealth of expertise to explain just what a fascinating field quantum physics is, Rae points out that it is not simply a maze of technical jargon and philosophical ideas, but a reality which affects our daily lives.

"Rae has done an impressive job. Any reader who is prepared to put in a little effort will come away from this book with an understanding of the basics of some important practical applications of the theory and some appreciation of why its conceptual foundations are still the subject of such spirited debate."
Professor Anthony Leggett – Winner of the 2003 Nobel Prize for Physics

"Rae's emphasis on the practical impact of abstract concepts is very welcome."
Professor Sir Michael Berry – Royal Society Research Fellow, Bristol University

ALASTAIR RAE is editor of *The European Journal of Physics* and was Reader in Quantum Physics at the University of Birmingham until his recent retirement.